Internet of Things and Fog Computing-Enabled Solutions for Real-Life Challenges

In today's world, the use of technology is growing rapidly, and people need effective solutions for their real-life problems. This book discusses smart applications of associated technologies to develop cohesive and comprehensive solutions for the betterment of humankind. It comprehensively covers the effective use of the Internet of Things (IoT), wireless sensor network, wearable sensors, body area network, cloud computing, and distributed computing methodologies. The book comprehensively covers IoT and fog computing sensor-supported technologies or protocols including web of things, near-field communication, 6LoWPAN, LoRAWAN, XMPP, DDS, LwM2M, Mesh Protocol, and radio-frequency identification.

The book

- Discusses smart applications to develop cohesive and comprehensive solutions for real-life problems.
- Covers analytical descriptions with appropriate simulation and prototype models.
- Examines the role of IoT and fog computing technologies during global emergency situations.
- Discusses key technologies including cloud computing, 5G communication, big data, artificial intelligence, control systems, and wearable sensors.

The text is primarily written for graduate students and academic researchers working in diverse fields of electrical engineering, biomedical engineering, electronics and communication engineering, computer engineering, and information technology.

Internet of Things and Fog Computing-Enabled Solutions for Real-Life Challenges

Edited by
Anil Saroliya, Ajay Rana, Vivek Kumar,
José Sebastián Gutiérrez Calderón, and
Senthil Athithan

Front cover image: Panchenko Vladimir/Shutterstock

First edition published 2023
by CRC Press
6000 Broken Sound Parkway NW, Suite 300, Boca Raton, FL 33487-2742

and by CRC Press
4 Park Square, Milton Park, Abingdon, Oxon, OX14 4RN

CRC Press is an imprint of Taylor & Francis Group, LLC

ISBN: 978-1-032-13631-8 (hbk)
ISBN: 978-1-032-13635-6 (pbk)
ISBN: 978-1-003-23023-6 (ebk)

DOI: 10.1201/9781003230236

Typeset in Sabon
by Newgen Publishing UK

Contents

Preface

As the use of technology is growing rapidly, people need effective solutions for their real-life problems. Engineering and scientific solutions somehow solve the problems of a common man in many diverse fields in an outstanding manner. However, the society is still facing a lot of troubles in the emergency situation like that of a "Pandemic". The term "Pandemic" is not new for the globe but dealing and handling the effects of the pandemic has always been challenging for us. Recently, the entire globe came into the grief of the COVID-19 pandemic, life got jammed, and everybody has become unstable except the pandemic warriors. But some challenges even persist for the pandemic warriors at a particular level due to lack in specific type of advance preparations; questions like "How this similar kind of situations can be handled effectually? How effectively resources can be managed? How can people live their life happily without much scuffle under such panic situation?" still arise in everybody's mind. This book mainly focuses on the solutions to the abovementioned questions by covering many areas that deal with the Internet of Things (IoT) and fog computing-enabled effective solutions for the society, amid the situation of emergency.

This book exhibits the extensive smart applications of associated technologies to develop cohesive and comprehensive solutions for real-life problems. The objective of this book is to explore all sorts of remedial technical options through which life of individuals can be unaffected in such critical situations. This can be easily done by reconnoitering the effective solutions with the use of sensor-enabled controlled systems. The chapters of this book mainly concentrate on the effective use of IoT, wireless sensor network, wearable sensors, Zigbee, body area network, cloud computing, distributed computing and fog computing methodologies so that similar opportunities and application challenges can be identified.

This book covers advanced applications of many related fields like AI-based IoT systems, effective use of sensors with human–computer interaction through high-speed internet, fog computing-oriented smart solutions, big data management for edge devices, and many more cutting-edge topics that can be used to make people's life easier during the phase of emergency. There is a lot of scope to achieve automation using IoT and fog computing in individual's life like road safety, smart hospitalization and medical care systems, goods and medicines transportation in remote areas, implementation of virtual safe environment for respective officials and social workers, etc. Technology has always made a person's life easier, comfortable, and safe from any emergency. With the help of these solutions, our society will be capable of dealing with such a panic situation easily.

About the editors

Anil Saroliya is working as an associate professor in Amity University Tashkent, Uzbekistan. His core research area is related to the Internet of Things (IoT), Edge Computing, Peer to Peer Networks, and Artificial Intelligence. He has published more than 35 research papers in reputed International Journals and Proceedings of International Conferences. He also has two patents in the field of IoT. He has conducted various seminars and workshops and delivered several guest lectures in the field of IoT and related areas. He was also a key member of e-Yantra Project of IIT Bombay and successfully established the Robotics lab under e-Yantra Lab Setup Initiative (eLSI) in Mody University of Science and Technology, Lakshmangarh, Sikar. He had also established IoT and Drone lab in Mody University campus. It is now helping the students and faculty members in their research and development-related experiments. Saroliya was also the founder member of RAIOT Lab (Robotics, Automation and IOT Lab) in Amity University, Rajasthan.

Ajay Rana is a known name amongst the Amity fraternity as he has been one of the founding members of many initiatives here at Amity. For almost 2 decades he has held various positions at Amity Education Group including the Director for Amity Institute of Information Technology, Group Director for Amity School of Engineering – JEE, Director for Amity Technical Placement Centre, Dean, Sr. Vice President & Advisor to the Amity Education Group for multiple roles. He has also served Shobhit University, Uttar Pradesh in the capacity of Vice Chancellor for some time.

Dr. Rana possesses a wide understanding of the acts, statutes, and ordinances of the University. Well, read about the NEP and expert in the Curricula development as per the NEP guidance. He is currently also a Member of the Consultative Group NEP 2020 CBSE, Ministry of Education, GoI, and Independent Director of Rajasthan Venture Capital Fund (GoR), Rajasthan Trustee Company Private Limited, GoR, India, and Sehaj Synergy Technologies Private Limited, India.

He has worked closely with the examination system and carries experience in carrying out various rankings, accreditations, and affiliations such as UGC, NAAC, NIRF, QAA, QS, WASC, WSCUC, and IET and statutory Bodies like AICTE, NCT, and BCI.

He possesses an outstanding academic record, a product of a prestigious system throughout. Obtained M.Tech and Ph.D. in Computer Science and Engineering. He has been conferred with Honorary Professorship from various Universities.

His areas of interest include Machine Learning, the Internet of Things (IoT), Augmented Reality, Software Engineering, and Soft Computing. He has 72 plus patents under his name in the field of IoT, Networks, and Sensors. He has published more than 279 Research Papers in reputed Journals like ACM, Springer, Elsevier, Taylor and Francis, and others, and International and National Conferences, Co-authored 09 Books and co-edited 36 Conference Proceedings.

Prof Rana is Founding Chairman of AUN Research Labs, EC Member of the IEEE UP Section, Senior Member of IEEE, and Life Member of the CSI and ISTE. He is also a member of the Editorial Board and Review Committee of several Journals.

With an intent to provide a robust learning environment to students and to strengthen the Indian Education System, Dr. Rana has visited many Top Universities & Colleges which have a legacy of more than 1200 years of producing leaders around the globe. Dr. Rana has excellent networking and social skills and a vision for enhancing professional connections with Top Academicians, Industrialists, and Bureaucrats at both national and international levels.

He has undertaken 45 Sponsored Research Projects and 18 major systems-based Management Consultancies in several reputed organizations both public and private, in India and abroad. 18 students have completed their Ph.D. under him. He has organized more than 504 conferences, workshops, faculty development programs, Seminars, and Talks sponsored by IEEE, Springer, CSI, and others.

Dr. Ajay Rana has been honored with 243 awards and recognition for his extraordinary work in the field of Education and Research.

Dr. Rana is well-regarded for his deep commitment, indestructible determination, humbleness, and passion for societal upliftment Dr. Ajay Rana is an educationalist, a teacher, an innovator, and a strategist, and a committed philanthropist.

Dr. Rana possesses deep organizational ethics, and equality, and believes in holding the hands of every individual who wishes to succeed in life.

Vivek Kumar is the founder and Vice Chancellor of Quantum University, Roorkee. Kumar is a professor of Computer Science and has experience in industry, teaching, and administration. He has a master's degree in the field of Computer Engineering from BITS, Pilani, and a Doctorate in Engineering (Artificial Intelligence) from the Faculty of Engineering, Dayalbagh Educational Institute, Deemed University. Kumar is a Fellow of IETE, senior member of IEEE, ACM, CSI, and International Association of Engineers, USA. He is a member of the board of studies of many reputed universities. Vivek Kumar has 29 years of rich experience in academics, research, and consultancy. Since his joining at Quantum University, he has established strong foundation for university automation, practices and procedures in teaching learning, examination system, interdisciplinary education system, and industrial presence. He has been instrumental in industry tie-ups with IT giants like Xebia, Palo-Alto USA, Google, Salesforce, Oracle, and Certiport USA to make university courses more acceptable in industry during the last year. Before joining Quantum University as VC, he has worked as Director-Principal of Delhi College of Technology and Management for more than a decade. He started his career as a software engineer at Hexaware Limited. He has served SIT Mathura as Dean Academic and as head of the department at Lovely Institute of Technology, Jalandhar. He has served Galgotias College Engineering and Technology as "Professor" and Head of Information Technology and MCA Departments for eight years. He played a major inning in placing Galgotias to the national platform. As a professor, he has taught many IR 4.0 technologies like Neural Networks, Fuzzy Logics, Artificial Intelligence, Distributed Systems, Digital Image Processing, and Theoretical Computer Science around three decades. He has developed many intelligent and expert systems as a part of his research contribution. He has published more than 80 international research papers and has attended many conferences and seminars in India, United States, Europe, and Singapore as keynote speaker. He has filed eight patents. Vivek Kumar has guided more than 58 MTech and MPhil level projects. He has supervised eight PhD. Presently six PhD are being supervised under his guidance in MDU and YMCA university on cloud computing and cancer research.

Sebastián Gutiérrez is Associate Professor at the School of Engineering at San Sebastián (TECNUN-Universidad de Navarra). He received an engineering degree in Electronics from Universidad Bonaterra (Mexico) in 2003, an MSc degree in Robotics from Universidad Panamericana (Mexico) in 2005, and a PhD in Automation and Industrial Electronics from Universidad de Navarra (Spain) in 2012. He has participated in ten industrial projects. He has authored more than 60 journals and conference papers. His main research interests are Internet of Things, wireless sensor network, home automation, smart buildings, solar energy, mechatronics systems, and robotics. He is a member Level I of the Mexican National Systems of Researchers (SNI).

Senthil Athithan is working as Professor in the Department of Computer Science and Engineering, Koneru Lakshmaiah Education Foundation,

Vaddeswaram, Andhra Pradesh, India. He has done BE in CSE, MTech (CSE), and PhD (CSE). His primary research area is related to cellular automata, computational epidemiology, and Internet of Things. He has published more than 20 papers in reputed International Journals and Proceedings of International Conferences. He was also awarded with "Vishisht Sikshak Puraskar" (Best Faculty award) by SGI in 2007. He is a program evaluator of CAC-ABET, United States. He has also authored a book on "Optimized Landmine Detection using Cellular Automata", Lambert Academic Publishing-Germany in 2011. He is consistently striving to create a challenging and engaging learning environment in which students become life-long scholars and learners. He also accomplished programming skills in Brix-CC, Lex, YACC, EPISIM, Oracle, Access, Moodle, etc.

Contributors

Pooja Anand
Amity University, Greater Noida, Uttar Pradesh, India

M. Bhuvaneswari
Sri Krishna College of Engineering and Technology, Coimbatore, India

R. Arun Chakravarthy
KGiSL Institute of Technology, Coimbatore, India

Rajiv Dey
BML Munjal University, Gurugram, Haryana, India

Arvind Dhaka
Manipal University Jaipur, Jaipur, India

Kamlesh Gautam
Poornima College of Engineering, Jaipur, India

Abhishek Jain
ShriRam Institute of Information Technology, Morena, Madhya Pradesh, India

Vinay Jain
ShriRam College of Pharmacy, Morena, Madhya Pradesh, India

Adlin Jebakumari
Jain Deemed to be University, Jayanagar, Bengaluru, Karnataka, India

Amita Nandal
Manipal University Jaipur, Jaipur, India

Chitra Pandian
Thiagarajar College of Engineering, Madurai, India

Anushka Phougat
Mody University of Science and Technology, Lakshmangarh, Skar, Rajasthan, India

Muratova Sitora Rustamovna
Amity University in Tashkent, Uzbekistan

Anil Saroliya
Amity University, Tashkent, Uzbekistan

Abirami Sasinthiran
Vellore Institute of Technology, Chennai, India

S. Sasipriya
Sri Krishna College of Engineering and Technology, Coimbatore, India

Gautam Seervi
Arya Institute of Engineering Technology and Management, Jaipur, India

Arpit Kumar Sharma
Manipal University Jaipur, Jaipur, India

Mayank Sharma
Amity University Uttar Pradesh (AUUP), Noida, India

Pankaj Sharma
ShriRam College of Pharmacy, Morena, Madhya Pradesh, India

Shivendra Singh
Arya Institute of Engineering Technology and Management, Jaipur, India

Mukul Tailang
School of Studies in Pharmaceutical Sciences, Jiwaji University, Gwalior, Madhya Pradesh, India

Rajashree Taparia
Rajasthan Technical University, Kota, Rajasthan, India

Taskeen Zaidi
Jain Deemed to be University
Jayanagar, Bengaluru, Karnataka, India

Chapter 1

Fog computing fundamentals in the Internet of Things

A taxonomy, survey, and future directions

Pooja Anand, Anil Saroliya, and Mayank Sharma

CONTENTS

1.1 INTRODUCTION

In the coming centuries, there are several Internet of Things (IoT) devices have enlarged day by day. To address a few essential demanding situations

DOI: 10.1201/9781003230236-1

in structure design one of the famous techniques of physical separation of purposeful gadgets is in client–server architecture. The server side of this separation is hidden within the cloud infrastructure in the mode of a web-scale device. This version is serving a varied range of applications jogging over the Internet, providing storage, computing power, and redundant services for reliability. But within the new paradigm of the IoT, the old separation partially fails to meet the set of system requirements. To lead the computation-intensive demands of low latency and real-time programs of geographically distributed IoT devices, a brand new "fog computing" paradigm has been added. Commonly, fog computing is living toward the IoT gadgets and extends the cloud-based totally computing, storage, and network-based facilities. Fog computing is introduced as an intermediate layer between the infrastructure on cloud and the clients. It brings computing, storage, management, and Internet/network services among others. In this chapter, we discuss the functions of fog computing, its advantages, and internal information and present case research to illustrate it in real software eventualities. The challenges in fogs can be comprehensively analyzed as an intermediate layer among IoT gadgets/sensors and cloud data centers [3]. As per specific challenging situations and capabilities, the classification of fog computing can be presented. In addition, we assign current works to the rating to determine our present-day gaps studies within the periphery of fog computing. We recommend future instructions for studies on the basis of their observations. The goal of this chapter is to provide a higher-level overview of fog computing at a higher stage for real-life challenges.

1.2 WHAT IS INTERNET OF THINGS

Internet of factors refers to physical and digital item that have particular identities and are linked to the net. This allows the improvement of sensible packages that make strength, logistics, commercial management, retail, agriculture, and many other domain names of human enterprise smarter. IoT allows different types of devices, appliances, users, and machines to communicate and exchange data; the application of IoT spans a wide variety of areas like homes, cities, environment, energy systems, retail, logistics, industry, agriculture, and health. Things in IoT refer to IoT devices which have unique identities and allow remote sensing, actuating, and remote monitoring capabilities. Almost all IoT gadgets generate statistics in a few shapes or the alternative, which whilst continue via facts analytics results in useful facts to guide in addition movements.

- The *IoT* in other word refers to the ever-growing network of physical items that feature an IP address for Internet connectivity, and the

internetwork communication that happens between these gadgets and other net-enabled devices and systems.
- Communication is done between all the Internet-enabled devices and systems.
- In simple words, *IoT* is an ecosystem of connected physical objects.
- It is also referred as *Machine-to-Machine (M2M)*, *Skynet*, or *Internet of Everything*.
- The *IoT* is expected to become a global networking infrastructure for cyber-physical systems.

1.2.1 How IoT works

First, sensors or units accumulate records from their environment. These facts should be as easy as a temperature analyzing or as complicated as a full video feed. We use "sensors/devices," it is due to multiple sensors can be bundled collectively or sensors may be segment of devices. For example, your telephone is a system that has a couple of sensors (camera, accelerometer, GPS, etc.), but your telephone is no longer simply a sensor when you consider that it can additionally function many actions.

Next, those records are dispatched to the cloud; however, it wants a way to get there! The sensors/devices can connect to the cloud via a range of techniques including cellular, satellite, Wi-Fi, Bluetooth, low-power wide-area networks [8], connecting through a gateway/router or connecting immediately to the net by means of ethernet (don't worry, we'll provide an explanation for greater about what these all imply in our connectivity section). Each alternative has trade-offs between strength consumption, range, and bandwidth. Choosing which connectivity alternative is exceptional comes down to the unique IoT application; however, they all accomplish the identical task: getting information to the cloud. Once the facts receive to the cloud (we'll cowl what the cloud capacity in our facts processing section), software program performs some variety of processing on it. This may want to be very simple, such as checking that the temperature studying is inside an ideal range. Or it may want to additionally be very complex, such as the usage of laptop imaginative and prescient on video to perceive objects. Next, the facts are made beneficial to the end user in some way. This ought to be through an alert to the person (email, text, notification, etc.). For example, a textual content alert when the temperature is too excessive in the company's bloodless storage. A person would possibly have an interface that permits them to proactively test in on the system. For example, a person may prefer to test the video feeds on a number of homes by a telephone app or an Internet browser. However, it's now not usually a one-way street. Depending on the IoT application, the consumer may additionally be in a position to function a motion and have an effect on the system. For example, the consumer may remotely modify the temperature in the bloodless storage through an app on their phone [5].

1.2.2 IoT-enabling technology

- Low-strength embedded system: High performance and less battery consumption are the inverse factors that play a good-sized position throughout the design of digital systems.
- Cloud computing: Data accumulated via IoT devices is huge and this record needs to be saved on a reliable storage server. That is in which cloud computing comes into play. The facts are processed and discovered, giving extra room for us to find out where such things as electric faults/mistakes are in the machine.
- Availability of big data: We know that IoT relies closely on sensors, especially in actual time. As these digital gadgets unfold in the course of each field, their utilization is going to trigger a huge flux of huge records.
- Networking connection: In order to communicate, net connectivity is required in which every bodily object is represented by way of an Internet Protocol (IP) address. But there are most effective a confined wide variety of addresses to be had in keeping with the IP naming. Because of the growing quantity of devices, this naming system will now not be feasible anymore. Consequently, researchers are searching out another opportunity naming machine to represent every physical item.
- RFIDS: Wireless microchips are used for automated unique identification of something via labeling it over them. Due to the fact that interconnection of things is the primary area of IoT, the RFIDS labels become handshaken with IoT technology and are used to provide the particular [11] id for the linked "things" in IoT [5, 8]. It makes use of radio waves in order to track the tags electronically which is attached with every physical object.
- Sensors: A sensor is a device that measures physical reading from its surroundings and converts it into records that can be interpreted through either a human or a machine. Nanotechnology are extremely small devices with dimensions typically much less than a hundred nanometers. Smart sensors are the vital enablers of IoT (see Figure 1.1).

This happens because of the IoT technology behind it:

- The temperature sensor related with the plant pot detects the low temperature.

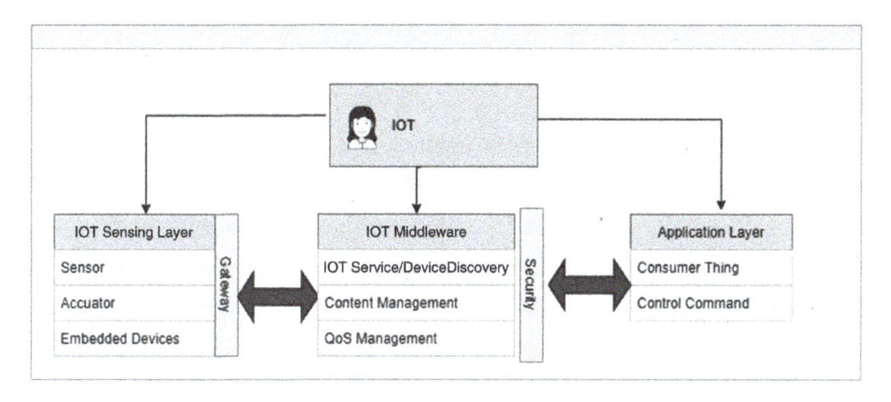

Figure 1.1 IoT architecture.

- Then, it triggers the microprocessor structures such as raspberry-pi and Arduino boards.
- It receives the sensor alerts thru Internet pathways including Wi-Fi and Bluetooth.
- Then, it notifies the human and the movement sensor linked to the tap which activates to pour it.
- **Actuators** Devices work in a machine to perform a mechanical action. It transforms electrical alerts into bodily actions. Each sensor and actuator works like transducers that convert one shape of electricity to some other. The alternate of facts is the maximum critical key factor in IoT. Consequently, sensors and actuators play essential function here.

1.3 CLOUD COMPUTING

Cloud computing [11] is a transformative computing paradigm that includes delivering packages and offerings over the net [6]. Cloud computing involves provisioning of computing, networking and storage sources on demand, and supplying those resources as metered services to the customers in a "pay as you pass" version. Cloud computing sources can be provisioned on call for via the customers in a "pay as you go" version. Cloud computing assets may be provisioned on call for by the customers, without requiring interactions with the cloud carrier company. The procedure of provisioning resources is automated [7]. Cloud computing resources may be accessed over the community using well known get entry to mechanism that offer platform-independent get right of entry to thru the use of heterogeneous consumer platform, which include workstations, laptops, tablets, and clever phones. The computing and storage sources provided by using cloud carrier

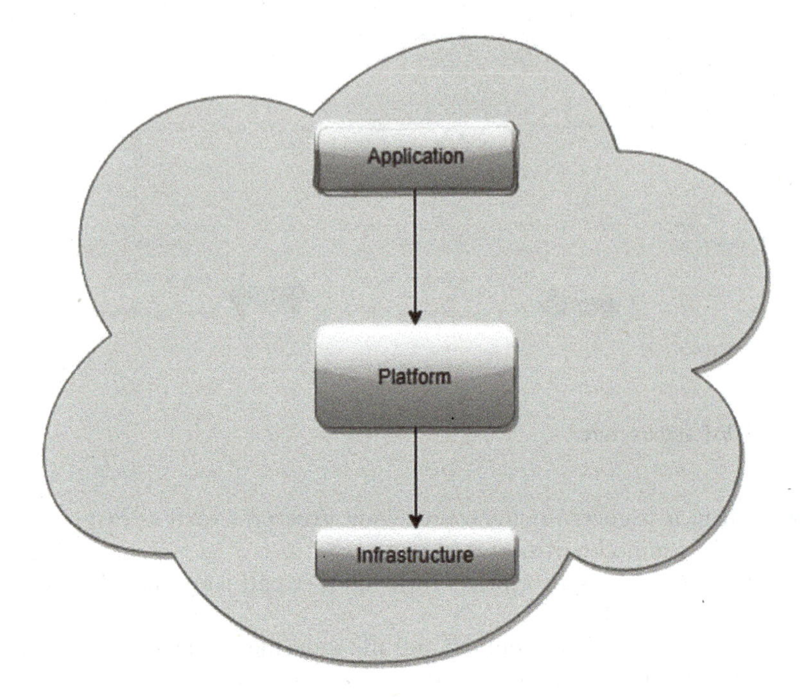

Figure 1.2 Layers of cloud computing.

providers are pooled to serve multiple customers the usage of multi-tenancy. Multiple customers can be served using multi-tenant components of the cloud via the equal physical hardware. Customers are assigned digital assets that run on the pinnacle of the bodily assets [8].

1.3.1 Cloud layers

There are three layers in cloud computing (Figure 1.2). These layers are used by companies on the basis of the service they provide.

- Infrastructure
- Platform
- Application

1.3.2 Cloud computing architecture

Cloud computing architecture is directly referred to the components and subcomponents (see Figure 1.3). These components directly refer to:

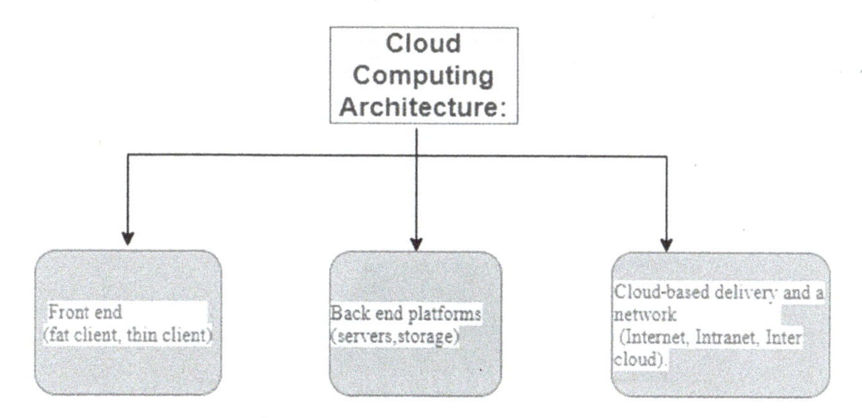

Figure 1.3 Cloud architecture.

1. Front end (fat client, thin client)
2. Back-end platforms (servers, storage)
3. Cloud-based delivery and a network (Internet, Intranet, Inter cloud).

1.3.3 Benefits of cloud computing

- Scalability: With cloud web hosting, it is simple to add and remove the number of servers based on the need. That is achieved by using either growing or reducing the assets within the cloud. This potential to adjust plans because of fluctuation in business length and desires is a wonderful advantage of cloud computing, specially whilst experiencing a sudden increase in demand.
- Instant: Anything you want is straight away available inside the cloud.
- Save money: A major benefit of cloud computing is in control on hardware fee. Alternatively of purchasing in-house system, hardware needs are left to the seller. For groups which are developing rapidly, new hardware may be a big, highly priced, and inconvenience. Cloud computing alleviates those issues because resources can be obtained speedy and without problems. Even better, the fee of repairing or replacing device is passed to the companies. Together with purchase value, off-website online hardware cuts internal strength costs and saves area. Big data facilities can take in treasured office space and produce a big quantity of warmth. Transferring to cloud packages or garages can help maximize space and notably reduce electricity prices.
- Reliability: In place of being hosted on one single time of a bodily server, hosting is brought on a digital partition which draws its useful

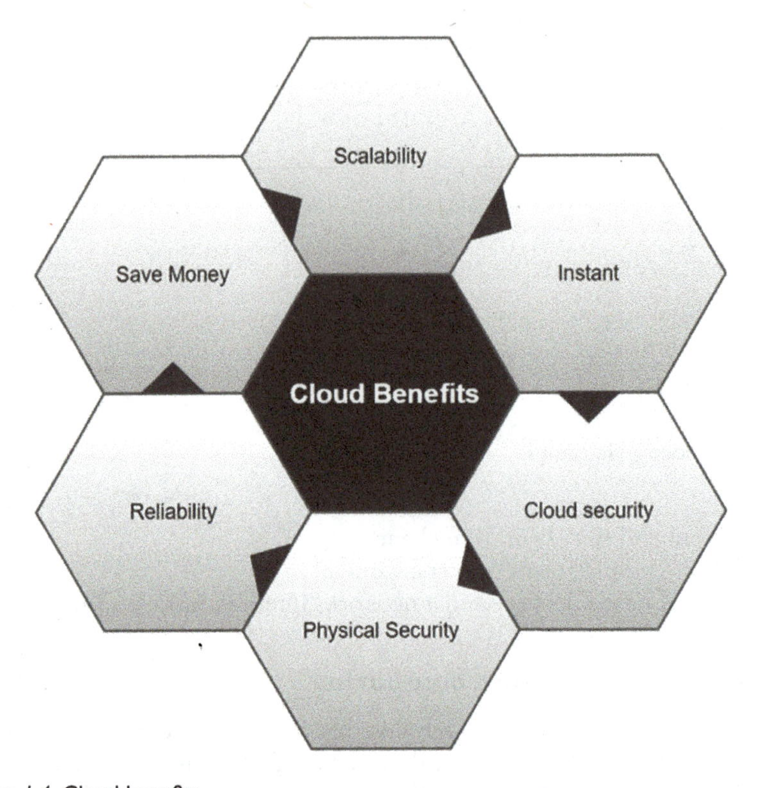

Figure 1.4 Cloud benefits.

resource, such as disk space, from an intensive network of underlying physical servers. If one server goes offline, it'll have no effect on availability, as the digital servers will retain tug aid from the ultimate community of servers.

- Physical security: The underlying physical servers are nevertheless housed within facts centers and so enjoy the security measures that those facilities enforce to save you humans having access to or disrupting them on web page detail view in Figure 1.4.

1.4 FOG COMPUTING

Fog computing explains the idea of cloud computing to the network edge, making it ideal for IoT and other applications that require real-time interactions.

Fog computing is a decentralized computing infrastructure in which information, compute, storage and applications are placed someplace among the information source and the cloud. Like side computing, fog computing brings the blessings and power of the cloud closer to wherein information is created and acted upon. Many humans use the terms fog computing and side computing interchangeably due to the fact both contain bringing intelligence and processing towards wherein the facts is created. That is regularly carried out to enhance performance, though it might also be performed for protection and compliance reasons. Fog is some other layer of a dispensed network environment and is intently linked with cloud computing and the net of factors (IoT). Public infrastructure as a carrier (IAAS) cloud companies can be idea of as a high-stage, worldwide endpoint for records; the advantage of the community is in which statistics from IoT devices is created. Fog computing is the concept of an allotted network that connects these environments. "FOG computing presents the lacking hyperlink for what facts desires to be driven to the cloud, and what may be analyzed domestically, at the brink," explains Mung Chiang, dean of Purdue College of Engineering and one of the country's pinnacle researchers on fog and side computing.

To guide the computational demand of real-time latency-sensitive programs of largely geo-distributed IoT devices [11], a brand-new computing paradigm named "fog computing" has been added [9]. The fog extends the cloud to be in the direction of the things that produce and act on IoT records. Those gadgets, referred to as fog nodes, may be deployed everywhere with a community connection: on a manufacturing unit ground, on pinnacle of an energy pole, alongside a railway track, in an automobile, or on an oil rig. Any device with computing, garage, and network connectivity can be a fog node. Examples include industrial controllers, switches, routers, embedded servers, and video surveillance cameras [4]. IDC approximations the number of facts analyzed on gadgets that are physically near the net of factors is drawing near 40% [1]. There is right purpose: analyzing IoT statistics close to wherein it is accrued minimizes latency. It offloads gigabytes of network visitors from the middle network, and it maintains touchy data in the community. Studying IoT records close to where it is accumulated minimizes latency. It offloads gigabytes of community traffic from the middle network. And it continues sensitive information in the community.

Fog programs are as numerous because of the Internet of factors itself. What they have got in common is tracking or studying actual-time statistics from network-linked things and then initiating an action. The movement can contain system-to-device (M2M) communications or human-device interaction. Examples consist of locking a door, changing device settings, making use of the brakes on a teach, zooming a video digicam, beginning a valve in response to a stress studying, creating a bar chart, or sending an alert to a technician to make a preventive repair. The possibilities are unlimited as shown in Figure 1.5.

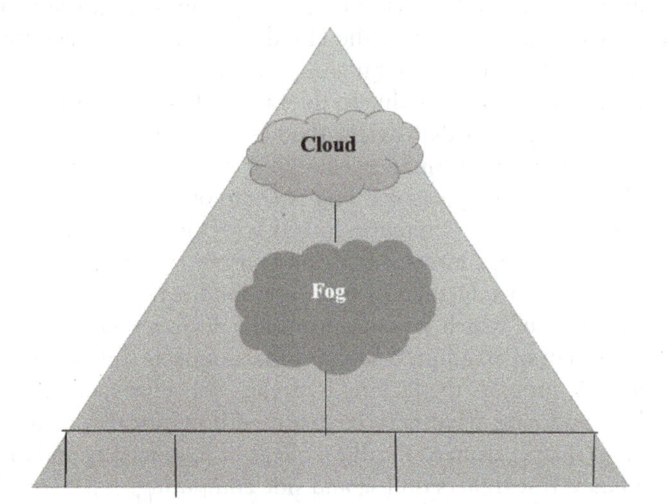

Figure 1.5 Fog computing architecture.

1.4.1 History of fog computing

In 2015, Cisco partnered with Microsoft, Dell, Intel, Arm, and Princeton University to form the Open Fog Consortium. Other organizations, including General Electric, Foxconn, and Hitachi, also contributed to this consortium. The consortium's primary goals were to both promote and standardize fog computing. The consortium merged with the Industrial Internet Consortium in 2019.

1.4.2 How fog computing works

Fog networking complements would not update cloud computing; fogging allows quick-term analytics at the edge, simultaneously the cloud plays useful resource-intensive, longer-term analytics. Although side gadgets and sensors are where records are generated and gathered, they don't have the compute and garage resources to perform superior analytics and machine getting to know responsibilities. Although cloud servers have the strength to do this, they are regularly too far away to system the records and respond in a well-timed manner. A fog computing material can have a ramification of components and capabilities. It can consist of fog computing gateways that accept records which is collected by IoT gadgets. It can include a selection of stressed-out and Wi-Fi granular series endpoints, which includes ruggedized routers and switching device. Different elements could include consumer premise system and gateways to get right of entry to area nodes. Higher up the stack fog computing architectures would also contact center networks and routers and finally worldwide cloud services and servers.

In addition, with that it having all endpoints connecting to and sending raw records to the cloud over the net may have privateness, safety, and legal implications, specifically while managing touchy facts difficulty to regulations in one-of-a-kind international locations. Popular fog computing packages encompass smart grids, smart towns, smart buildings, automobile networks, and software-defined networks.

1.4.3 Application of fog computing

There are a variety of abilities which use instances for fog computing. Unexceptional case for fog computing is traffic manipulation. Due to the fact, sensors – including the ones used to come across visitors – are often connected to mobile networks, cities on occasion deploy computing resources close to the cellular tower. These computing competencies permit actual-time analytics of traffic facts, thereby allowing visitors alerts to respond in actual time to converting situations.

This basic idea is also being extended to an autonomous vehicle. Autonomous vehicle basically functions as edge devices due to their substantial onboard computing power. Those motors are able to ingest records from a huge quantity of sensors and carry out real-time records analytics after which respond thus.

Due to the fact, an autonomous vehicle is designed in this way that does not need any cloud connectivity; it is tempting to think of automobiles as not being related gadgets. Even though an independent vehicle has been able to power properly inside the overall absence of cloud connectivity, it's nonetheless possible to use connectivity whilst to be had. A few cities are considering how an autonomous automobile would possibly operate with the identical computing resources used to govern visitors' lighting. This type of vehicle might, for example, feature as an edge tool and use its very own computing competencies to relay actual-time facts to the device that ingests visitors' records from other sources. Then underlying computing platforms use this information to function site visitor alerts more correctly.

1.4.4 Fog computing benefits and drawbacks

Like any other technology, fog computing has its pros and cons. Basically, the development of fog computing frameworks offers organizations greater alternatives for processing data anyplace it's far most suitable to do so. For some packages, information can also want to be processed as quickly as possible – as an example, in a production use case wherein linked machines need so one can respond to an incident as soon as possible.

Fog computing can create low-latency community connections between devices and analytics endpoints. This architecture in turn reduces the amount of bandwidth wished compared to if that information needed to be dispatched all of the way again to a statistics middle or cloud for processing. It may also be utilized in scenarios in which there is no bandwidth connection to ship records, so it has to be processed near in which it's far created. As an introduced gain, customers can location security features in a fog network, from segmented network traffic to virtual firewalls to protect it.

Some of the advantages of fog computing include the following:

- *Bandwidth conservation.* Fog computing reduces the volume of data that is sent to the cloud, thereby reducing bandwidth consumption and related costs.
- *Improved response time.* Because the initial data processing occurs near the data, latency is reduced, and overall responsiveness is improved. The goal is to provide millisecond-level responsiveness, enabling data to be handled in near real time.
- *Network-agnostic.* Although fog computing generally places compute resources at the LAN level – as opposed to the device level, which is the case with edge computing – the network could be considered as part of the fog computing architecture. All at once, though, fog computing is network-agnostic and the network can be wired, Wi-Fi or even 5G, as in Figure 1.6.

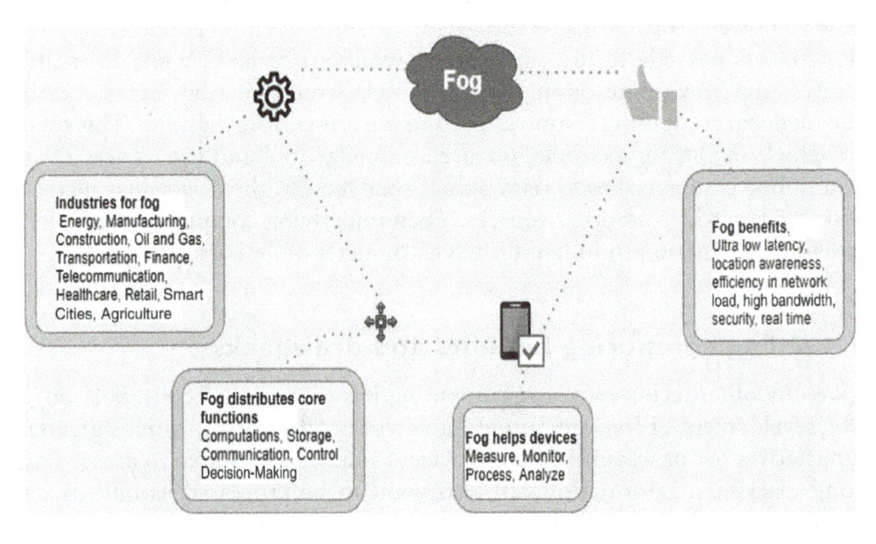

Figure 1.6 Benefits of fog computing.

Of course, fog computing also has its disadvantages, some of which include the following:

- *Physical location.* Because fog computing is tied to a physical location, it undermines some of the "anytime/anywhere" benefits associated with cloud computing.
- *Potential security issues.* Under the right circumstances, fog computing can be subject to security issues, such as IP address spoofing or man in the middle attacks.
- *Startup costs.* Fog computing is a solution that utilizes both edge and cloud resources; it means that hardware costs are associated with that.
- *Ambiguous concept.* Although fog computing has been about for several years, there is still some ambiguity around the definition of fog computing with various vendors defining fog computing differently.

1.4.5 What materializes inside the nodes of fog and the cloud?

- In actual time, it receives signal from IoT gadgets by the use of any protocol
- For actual-time manage and analytics IoT-enabled packages run, with millisecond reaction time
- Deliver brief garage, regularly for a maximum of two hours
- Send regular information summaries to the cloud platform
- Collect statistics summaries from many fog nodes
- Plays evaluation at the IoT records and records from other resources to gain enterprise perception
- New software guidelines can be sent to the fog nodes primarily based on those insights

1.4.6 IoT and fog computing

Because cloud computing isn't always viable for lots of Internet of factors (IoT) applications, fog computing is frequently used. Its disbursed technique addresses the wishes of IoT and industrial IoT, along with the titanic number of statistics clever sensors and IoT devices generate [8], which would be high-priced and time-ingesting to ship to the cloud for processing and analysis [10]. Fog computing reduces the bandwidth wished and decreases the lower back-and-forth communication between sensors and the cloud, which may negatively affect IoT overall performance.

The IoT is producing an extraordinary volume and variety of statistics (data). However, by the time the information makes its manner to the cloud for evaluation, the possibility to behave on it might be gone [8, 10]. This white paper, intended for it and operational technology experts, explains a new model for studying and acting on IoT information. Its miles known as either edge computing or fog computing:

- Analyzes the maximum time-sensitive information on the network area, close to wherein it is generated in place of sending tremendous quantities of IoT data to the cloud.
- Acts on IoT records in milliseconds, based totally on policy.
- Sends decided data to the cloud for early investigation and longer term garage.

These days cloud model isn't designed for the volume, range, and velocity of information that the IoT creates.

1.4.7 How IoT actually relates to your enterprise

The IoT quickens awareness and response to instances. In industries such as manufacturing, oil and fuel, utilities, transportation, mining, and the public region, quicker reaction time can enhance result, surge carrier levels, and rise protection.

These days, when business firms and businesses are facing colossal opposition, other challenges are just underneath the floor. If you combine any linked gadgets into your enterprise technique, you aren't going to take away them. But you can encounter some new vulnerabilities and demanding situations that may restrict your capacity to satisfy the end user's needs. The truth of the problem is that you could triumph over those issues as long as you implement IoT with severe care [8]. You cannot put off all the demanding situations at once. It calls for quite a few precautionary measures to reduce vulnerability. In commercial enterprise, one has to go through various challenges and limitations to achieve your dreams – this is the best manner to be successful.

You probably questioning that if there are such a lot of demanding situations to enforcing IoT, then what's the factor in the use of it at all? But the element is you're bound with the customers or customers' expectations. Their interest level is excessive, and that is why you want to accomplish that. However, there are methods, which I have recommended in this newsletter, that help you preserve IoT structures to manipulate. Therefore, you may face some common demanding situations so that you can be well prepared for your business enterprise.

1.4.8 How IoT actually relates to your infrastructure

Take advantage of the IoT needs a brand new form of setup. These days cloud model isn't designed for the volume, range, and velocity of information that the IoT creates. There are millions of previously distinct gadgets that are causing more and more data every day. An expected 50 billion "matters" could be related to the Internet by means of 2020. All statistics from these gadgets to the cloud for study could need huge quantities of bandwidth.

1.5 HOW TO TRACE AND SAVE YOU IN COVID-19 NETWORK TRANSMISSION WITH THE AID OF FOG COMPUTING

To slow the spread of COVID-19, governments across the world are trying to identify affected population and control the spread of virus using isolation and quarantine [2]. Because it's extremely difficult to hint those who have been in touch with an infected person, ensuing in huge community transmission and higher contamination costs. To address the issue, we developed PPMF, an e-authorities privacy-preserving cellular and fog computing platform that can detect inflamed and suspected cases across the country. To track community transmission while retaining user data privacy, we use nonpublic cell phones with apps to trace the contacts and multiple stationary fog nodes, such as automated risk checkers (ARC) and Suspected User Records Uploader Node [2] (SUDUN). A Unique Encrypted Reference Code [2] (UERC) is received by every consumer's cell device while enrolling at the crucial utility. Rotational Unique Encrypted Reference Code (RUERC) is generated by both, the urgent utility and the cellular tool, which is broadcast over Bluetooth. Since the arcs are placed at home entry points, they can continuously detect if there are suspected cases nearby [2]. In case if displayed cases are found, the arcs immediately start broadcasting warning messages to humans in range keeping the identity of the enraged character completely discrete [2]. Installation of the SUDUNS in health facilities is done to transfer the test results to the critical cloud utility. The data is further used to create a map of cases which are either suspected or infected. Using the PPMF framework, economic activities can be allowed by governments to groups without total lockdown. In 2019, ailment of the novel coronavirus (COVID-19) has spread rapidly in the whole world within a short time period. A massive crisis among population concerning their health was caused around the world; the spread was so deadly that within eight months of its first infection detection [2], it had killed more than 20 people. Cell apps can help alert each inflamed and suspected case in near real time, and governments are racing to research, develop, and make live such programs and frameworks. However, a lot of applications exacerbate major security and data issues concerning the privacy of users by collecting sensitive and, in my opinion, identifiable records of customers and failing to adhere to the transparency of user data ingestion and processing. Temporary lockdowns in cities were imposed by government to control the infection rate of COVID-19, resulting in massive financial setbacks. We can reduce the impact on economy by avoiding lockdowns on bigger scale as well as more precise isolations. As a result, implementing fog computing in public and economic zones (e.g., malls, agency homes) can ensure ongoing financial sports by alerting nearby people, whereas mobile computing (cellular apps) can assist in tracing the cases to reduce infection and suspected cases can be identified.

1.5.1 Contextual and problems

Q1 What are the problems and confidentiality alarm in existing connection tracing apps?
Solution: First, we must look into the heritage and issues of current software model along with user facts privacy concerns. There are numerous cellular software models developed by governments and other collaborators to track the transmission of the COVID-19 network. However, the majority of these packages and frameworks are unsuccessful in ensuring consumer data privacy and are plagued by a range of problems, such as compulsory app use, huge personal data accumulation, compromising transparency of source code and statistics drift, ineffective records storage and processing, and are unable to provide users right to delete their data.

Q2 How to utilize cellular and fog computing to detect and prevent you COVID-19 community transmission?
Solution: By disseminating the workflow, design issues, and structure of e-authorities utility framework, which makes use of mobile applications and the fog computing. There are two types of fog nodes (ARC and SUDUN) included in the system: a plethora of the RESTful APIs and a critical utility. Customers can use the person API to sign in themselves. The use of arc is to assess the risk of customers visiting commonly accessible places (e.g., mall, enterprise constructing). The check results of COVID-19 are sent to the main utility via the end result API by the check facilities. If the test results are excellent, uploading of regionally saved touch tracing statistics to cloud is done by individuals using the SUDUN or immediately from the cell app.

Q3 How to create an automation-driven privacy-preserving e-government model?
Solution: By discussing overview of the implementation and safety of consumer statistics in our framework. In this section, we will go over the development of a framework based on cloud, primarily Amazon Web Services (AWS). We provide privacy information and safety primarily on the basis of user consent (voluntary, compliance, and individual consent), nominal data collection (number, zip code, and age range), information deletion at the consumer's request, transparency (open-supply codes, clean records float), and restricted future use of records.

1.5.2 Scope

We expect our framework to be handing a population data, being deployed and managed by efficient and extensive government of a country. Governments have got right to entry to the test outcomes and can switch such an included cellular-fog computing context. Furthermore, counting on a nonpublic unit to manipulate this type of framework can restrict the upkeep of statistics privacy. Moreover, on these paintings, we in most cases recollect and pay attention to user information privacy problems. Regardless of whether or not we are developing a preferred and secure records processing framework, we wish not to discuss advanced safety threats concerned to the cell, fog, and cloud layer because there is a wealth of literature on gift safety functions.

System specifications

In our included framework, we present the following mobile software program and fog node functions:

1. Touch tracing: When registering a device, the unique reference code which is hashed is received by the user. A new specific reference code is generated every hour in the utility; the reference code is further shared with nearby devices with the user's permission. This process ensures that the statistics which is published cannot be used to track down a user. To access the danger level without disclosing the identity, similar hash could be generated within the cloud utility.
2. Self-checking: Using the mobile software program, one can determine if they were in close proximity to any infected victim within the previous 14 days. The software will upload the device's regionally saved reference codes from the previous 14 days. The cloud software monitor to see if a match between uploaded RUERC is associated with an infected or suspected patient. This can be used for notification to any individual.
3. Minimum cell computation: The proposed framework ensures minimum cell resource consumption by allowing consumers to give permissions or revoke access to whether scanning and background services can be turned on or off. Along with it, the mobile app has the intelligence to postpone broadcast. Self-testing and importing records will be the only cases with compulsory Internet requirements.
4. Individual private records are not saved and are abstracted from the machine, ensuring consumer privacy. We keep data requirements to a minimum. Any inflamed sufferer is lost in the fog of fog nodes.
5. Signals from fog nodes: To limit network transmission, we implement an automatic chance checker in public places such as shopping malls and workplaces. Because of the risk of user identity, a time delay is implemented along with limited human access before sending broadcast of simple alert messages. Those should only advise customers to take precautions and no longer display any RUERC or risk radius.

- System elements

Our framework is comprised of four primary additives: a mobile application that declares its RUERC and stores obtained RUERCs from nearby devices. In employer creation, the arc assesses inflamed cases and broadcasts alerts. The SUDUN uploads data from infected cases' devices. Finally, a critical cloud application integrates several of these cell and fog nodes and manages accumulated data.

1. Mobile application

Customers download the application from the google play or government server and install it in the cellular device. The app declares its RUERC and gets RUERC of various gadgets round. It additionally detects the UERC broadcasted, which is predefined, through automatic hazard checker (ARC) and alerts the client about suspected case spherical.

2. Automatic risk checker (IoT/fog) (ARC)

Our fog tool is available in medical facilities, purchasing centers, educational institutions, and all government-supervised businesses where network transmission is possible. The fog can collect the RUERC of Bluetooth-enabled mobile devices and connect and communicate with the services on cloud to locate any inflamed or suspected patient. If any RUERC is determined to be an enraged or suspected victim, it announces a specific UERC at some point in the region. The customers are alerted of the threat once the mobile application detects this UERC near the fog. Due to being registered with provider of cloud fleet, monitoring of arc node is possible in real time. As a result, high rate of COVID-19 affected population can be identified by arc nodes. If the frequency of infected or suspected patients exceeds a predefined threshold rate per hour, it continuously generates an emergency alert to the employer's authority through emails and messages. It alerts the fitness government which is present in the nearest range. Only at a sample size of 5, broadcasting of notification happens for ensuring the privacy of the inflamed victims. Hence, the immediate dispatch of notification is ignored. To ensure the privacy, the notification is sent only after a certain period of time.

3. Suspected user data uploader node (IoT/Fog) (SUDUN):

The fitness centers or COVID-19 checkpoints hold the position of the fog nodes which are similar to arc. When an examination center criticizes a superb examination outcome, the agitated character can allow the application to import its RUERC list from the privacy dashboard and keep the cellular phone in the form of an SUDUN. The stored list of hashed RUERC is fetched when the SUDUN routinely connects to the user's device. Monitoring in real time is allowed for each SUDUN through the use of a cloud dashboard. Furthermore, the infected patient can add the RUERC list without the assistance of an SUDUN by utilizing the cell software and a web connection.

4. Cloud/server

Facilities for secure data storage and computing servers are provided by the government. Its primary responsibilities include fetching and saving data to and from SUDUN, prediction of probable spreaders can be done by data mining, computation to provide required information to ARC, and standard authentication is handled to maintain security and privacy.

System workflow

1. User registration

Customers' registration required a combination of their mobile number, age group, and postal code to ensure uniqueness of identity. The cellular variety discipline is required, but the age group and postal code are optional. One-time password (OTP) is sent via SMS to the user-supplied cellular range. The customer then enters the unique OTP into the application, and the app transfers the OTP to the cloud via REST API. The cloud software program compares the OTP sent by the consumer to the originally created one. If all of the conditions are met, the cloud software generates a UERC and sends it back to the consumer's linked cellular app. Finally, the cell app saves the UERC in an encrypted format on the user's device before scanning ids broadcasted by other customers. A rotational UERC is pronounced by the person tool itself.

2. Key generation

A RUERC is generated by the model. If any user spends at least one hour near an agitated person, he becomes a suspect in the case. Because the RUERC changes every hour, customers may achieve multiple RUERC if they live nearby for that duration (e.g., if they're friends). Our app ensures that regardless of whether or not a person is nearby at any time, it will save the person's key and compare it to the timestamp. It improves contact location accuracy without requiring a rigorous computation. Furthermore, the system improves privacy by avoiding continuous broadcast of the same UERC, as it can result in the wireless monitoring. Once the UERC from the cloud is received by the application after registration, and from this UERC, using the AES RUERCs are generated every two hours by the c application language for 14 days period. Simultaneously they transfer and store these RUERCs.

3. RUERC scan service

Software execution is enabled by the check provider in the past. A thread is forked to mitigate the possibilities of difficulties by the client while using various cell apps. The character has the ability to stop and restart walking the carrier at any time. Its primary responsibility is to scan and market RUERC, grow documents, and save records. To exchange RUERC between two cell phones, scan provider employs Bluetooth Low Power (BLE) technology. BLE uses significantly less energy to communicate with, manipulate, and for monitoring of the IoT devices. It employs the terms "central" and "peripherals," which define a network known as a piconet. Primary scans for the industrial and peripheral manufacturers industrial. The commercial in our framework consists of the described RUERC. General characteristic profile (GATT) is designed to transfer and obtain brief bits of information called "attributes," it is based on the attribute protocol (ATT), which employs a 128-bit unique identifier. We are omitting the use of GATT and ATT in our framework to maintain the bare minimum of data collection insurance and device strength. We use peripherals to transmit RUERCs and critical to validate those RUERCs. Along with scanning, we remove the RUERCs that have been flagged as suspicious by algorithm to filter suspects and keep them for this reason.

4. RSSI

Our framework makes use of the received signal strength indicator (RSSI) to determine the distance between devices. RSSI denotes the strength of the obtained signal. It is calculated in decibels and is affected by the broadcasting device's strength and chipset. It is also affected by the medium of transmission. If there are any obstacles between the receiver and the sender, the signal energy and thus the RSSI will decrease. As a result, the rate of RSSI in a specific medium indicates distinct values for a one-of-a-kind producer. However, for a specific producer, RSSI indicates the depth of signal strength. RSSI no longer provides the correct distance; however, an approximation of distance can be found using the path loss model.

$$RSSI = -10 \cdot n \cdot \log10(d) + C$$

Here, n is the path loss exponent that depends on the transmitting medium, d is the distance, and C is a constant.

5. RUERC storage in mobile device

To protect the privacy of the user, the mobile software creates two documents within the cellular device's internal garage. Also, the access to these files is abstracted from the user. On receiving the signal by the peripheral, the signal facts are stored in one record temporarily, and another stores the filtered sign facts. Upon entering within the range of SUDUN by the tool, the signal records of the filtered report are sent to SUDUN or immediately to the cloud to perform self-test.

6. Suspect filtering

According to the Centers for Disease Control (CDC), a distance of six feet between the case and the point of contact is safe to keep. Someone is considered a touch if he or she is within six feet for at least 15 minutes or is within an hour's proximity. The experiment provider scans all of the RUERCs within Bluetooth's range. However, given the preceding rules, continuous scanning is not recommended because it depletes the battery and memory. As a result, the scanning carrier scans for 700 ms using a three-minute c programming language. We broaden our suspect filtering algorithm to include these elements in order to consume less energy and memory while identifying suspects more accurately. The mobile tool contains a variety of documents. On each c program language period, one includes all of the sign statistics (RUERC, distance, and timestamp) it receives from nearby devices.

1.5.2.1 Implementation overview

We built our framework with AWS. Because it is a widely used framework, it can be implemented in a variety of IoT/cloud structures. The fog nodes, arc, and SUDUN include AWS green grass and lambda components. They could be related to the use of the MQ telemetry delivery (MQTT) protocol by the AWS IoT core provider. The use of IoT tool defender ensures the authenticity and protection of the fog nodes. To display and audit the fog nodes, the IoT tool manage provider is used. Incoming data from fog nodes is routed to an easy queue carrier (sq.). The queue records are processed using lambda capabilities and moved into software containers, which are managed by the Kubernetes cluster within an auto-scaling institution for dynamic scaling of the master nodes and worker nodes. A Secure Shell (SSH) connection to a bastion server is provided for cloud application control via elastic load balancing. Customers want to be connected to the cloud software program only during the registration process via an https connection to a load balancer. The examination facilities deliver the examination outcomes through the use of a RESTful API over https. As an example, the scheduled responsibilities of producing the user's rotational UERC are completed in an elastic container carrier that uses the cloud watch event rule at a specific time of day. The application generates alarms for any routine activities, such as device connection errors or access failures, and sends them to the appropriate authority via cloud watch alarms. Customers receive notifications from the arc, SUDUN, and cloud packages via the easy notification provider. The simple email issuer is used for similarly conversation procedures to authorized groups and directors.

1.6 CONCLUSION

In this chapter, a bastion server was provided SSH connection for cloud application control via elastic load balancing. Customers prefer https connection with a load balancer to connect with the cloud software program during the registration. The examination facilities deliver the examination outcomes using a restful web service (API) over the https protocol. As an example, the scheduled responsibilities of producing the user's rotational UERC are completed using software program within an elastic container, time scheduled cloud watch event rules are used by the elastic. An alarm is generated by the application for routine activities, such as connection errors or access failures between devices, and sends them to the appropriate authority via cloud watch alarms. Customers receive notifications from SUDUN, the arc, and the cloud packages via the easy notification provider. The simple email issuer is used for similarly conversation procedures to authorized groups and directors. As a result, by utilizing the shape of PPMF framework, as proposed, governments will maintain monetary activities along with tracing and reducing mass-stage community transmission. Based on our framework, the intention is to broaden an excellent model for spreader detection and clustering technique of inflamed instances in the future.

REFERENCES

1. https://internetofthingsagenda.techtarget.com/definition/fog-computing-fogging
2. A Privacy-Preserving Mobile and Fog Computing Framework to Trace and Prevent COVID-19 Community Transmission | IEEE Journals & Magazine | IEEE Xplore.
3. 05 – oprea, tudorica, belciu, botha.pdf (ase.ro).
4. Simona OPREA1, Bogdan George TUDORICA2, Anda BELCIU1, Iuliana BOTHA1, Internet of Things, Challenges for Demand Side Management, Informatică Economică vol. 21, no. 4/2017.
5. www.ibm.com/in-en/cloud/learn/cloud-computing
6. www.geeksforgeeks.org/fog-computing/
7. www.geeksforgeeks.org/difference-between-cloud-computing-and-fog-computing/

8. Arshdeep Bahga and Madisetti Vijay, Internet of Things, A Hands-On Approach Paperback – 1 January 2015.
9. www.springer.com/series/11636
10. www.researchgate.net/publication/344370756_A_Privacy
11. Redowan Mahmud, Ramamohanarao Kotagiri, and Rajkumar Buyya, Fog Computing: A Taxonomy, Survey and Future Directions, Internet of Everything, 103–30, October 17, 2017. https://doi.org/10.1007/978-981-10-5861-5_5

Chapter 2

Role of IoT to control the movement of infections during pandemic

Muratova Sitora Rustamovna and Anil Saroliya

CONTENTS

2.1 INTRODUCTION

Throughout film history, many films have been shot about pandemics and epidemics, which ultimately led the whole world to chaos and devastation. This format of films is quite popular and often earns a good box office. Global catastrophes and terrible diseases appearing literally out of nowhere; attacks by aliens or fearsome mutants grown in secret laboratories – all this pretty well excites the imagination as well as gives free rein to fantasy. However, nobody would have guessed that someday something like this could happen in real life that the lives of millions of people from all over the world could be in mortal danger.

On December 31, 2019, the first outbreak of a new, previously unknown infection was reported in the Chinese city of Wuhan (Hubei province). This infection turned out to be a dangerous disease caused by a new coronavirus called SARS-CoV-2. In a matter of weeks, the COVID-19 virus covered almost all provinces of China and, without slowing down, began to spread around the world through tourists and air travel. COVID-19 is transmitted by airborne droplets or through touch, that is, you can be infected through door handles or handrails in the metro and buses, which have been touched

DOI: 10.1201/9781003230236-2

by infected people. Thus, the number of infected individuals rose at an astronomical rate, strict quarantine was declared all over the world, and the epidemic turned into a pandemic that affected absolutely everyone.

In order to fight the virus, it was necessary to prevent it or, at least, slow down its spread. However, since the virus is very contagious, and the incubation period is about two weeks, besides, do not forget about the group of people in whom the COVID-19 disease is completely asymptomatic; it will be extremely problematic to track all foci of spread. Exactly here a wide range of technologies comes to the aid of all humankind. The variety of different devices, gadgets, applications, and information networks that they form in aggregate helped to transfer our daily life to the online format quite easily and without significant losses due to the establishment of a complete quarantine. This, in turn, led to the need to "improve and wider use of the Internet of Things".

2.2 THE INTERNET OF THINGS

The Internet of Things (IoT) is a well-designed network for transmitting, analyzing, and processing data, consisting of various digital and mechanical devices. In simple words, these are devices connected to the Internet and to each other. Within the IoT, people can communicate with "things" and "things" with people. IoT systems usually consist of a network of smart devices and a cloud platform to which they are connected using various types of communication, such as Wi-Fi or Bluetooth. They are working in real time and devices function independently, they do not need total control, however, from time to time, people must adjust the operation of the network and provide additional access to data. Many industries are using IoT to understand consumer needs in real time, improve agility, immediately contribute to the improvement of the machines and systems, streamline operations, and find innovative ways to work as part of their digital transformation efforts.

The IoT system consists of only four main building blocks (Figure 2.1) – sensors, processors, gateways, and applications. Sensors are devices, which can collect and transmit information, because of this they are also called "things" of the IoT. They can be smart bracelets, gas safety sensors, or wearable sensors that read biometric data. Processors, on the other hand, process and manipulate data received from sensors, besides that they are responsible for encrypting and decrypting information, which means they are in charge of data security. Gateways are responsible for building routes for sending and receiving information; examples of gateways are LAN and WAN. Applications receive processed information, with which users subsequently work. Together, these blocks form a well-designed and useful network of the IoT.

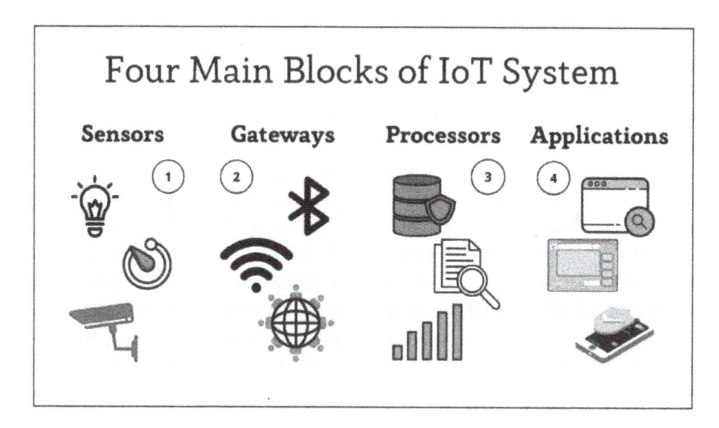

Figure 2.1 Main blocks of IoT system: sensors, gateways, processors, and applications.

IoT is commonly used in all spheres of human life. For example, in the power industry, the IoT improves the controllability of substations and power lines through remote monitoring. In agriculture, "smart" farms and greenhouses doze fertilizers and water themselves, and "smart" animal trackers notify farmers in time not only about the location of animals, but also about their health status, analyzing heart rate, body temperature, and general activity. In logistics, IoT reduces the cost of transportation and minimizes the impact of the human factor; IoT systems can also monitor the filling of garbage cans and optimize the cost of garbage collection based on these data. However, in 2020 much attention is paid to IoT in healthcare, because the health of the nation in the context of a global pandemic is a paramount task. IoT helps to monitor the state of equipment in wards and monitor the condition of critically ill patients using special sensors. Therefore, such devices are able to track and monitor the health of patients from the moment they arrive at the clinic, collect, and update data about them without involving junior medical personnel (which saves time and resources).

Furthermore, forced social distancing and complete isolation have led to the rapid development of technologies to help control the epidemiological situation in regions and countries. Tracking potential outbreaks, controlling the spread of the virus, quarantining, and treating those who were infected – all this would be impossible without the use of the IoT. For example, drones, which help monitor wearing masks and social distancing in public places, or special applications that help medical staff track the location of patients under quarantine in their homes.

2.3 CONTROLLING SOCIAL DISTANCES USING IOT

Since the coronavirus is mutating more and more and does not even think to slow down the rate of spread, people need to restore their usual way of life: start going to work, schools and universities again; safely face-to-face communicate with colleagues. Many companies and institutions are already using technologies that help quickly identify first signs of the disease. Temperature sensors and distance tracking applications are examples of such technologies. However, the threat of reinfection is constantly present, and only 60% of the sick can calculate temperature measurements, because some people are asymptomatic carriers of the virus, so it was decided to find new ways to control the spread as well as improve existing technologies [1].

The simplest way to protect yourself from the coronavirus is social distancing (Figure 2.2). It prevents the virus from spreading by airborne droplets, which can help to reduce the rate of infection of the population without setting strict quarantine. The minimum social distance between two people is two meters (six feet). On the one hand, this method of protection is quite simple and straightforward, but it is quite hard to comply with it. We are all used to communicating freely, without restrictions, and when new rules suddenly appear, it takes time to get used to them. Thus, in order to monitor compliance with the new rules and remind people of them, various devices of IoT systems began to be used: existing gadgets were modernized and completely new ones were invented.

To the conclusion that "all ingenious is simple" came one group of Dutch developers who launched their startup called Safezone [2]. Safezone is a distance meter that immediately warns people if they get too close to each other, helping to maintain a safe distance. This technology uses an ultraprecise ultrasonic sensor to measure the distance between people. If the distance is violated, a red light signal turns on; if people are at a sufficient

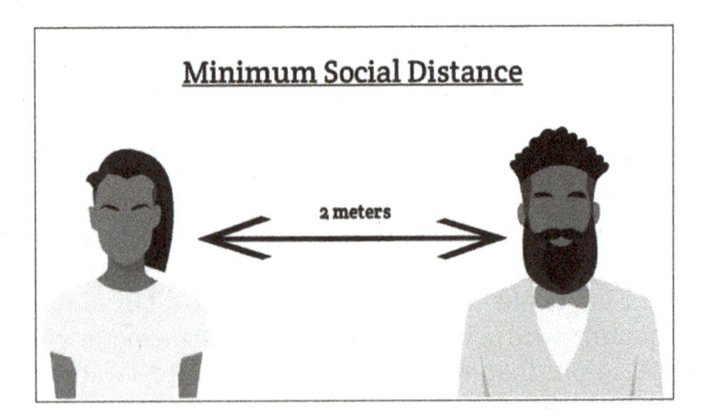

Figure 2.2 Two people stand at a distance of two meters from each other.

distance, then the signal lights up green. Moreover, you can optionally add an additional sound signal, or turn it off. Safezone helps ensure secure communication between employees of different companies, their owners and even guests.

Monitoring the observance of the distance can be carried out not only by the so-called distance meters, but also by applications on smartphones, and portable sensors that can be worn like bracelets or cling to clothes. Moreover, the presence of advanced software will allow you to identify places with a significant number of infected and build safer routes; it is possible to use the function of recording the time of contact with people for a more detailed analysis of the data (if, of course, the user himself allows it). Also, such monitoring of the movements and contacts of people with each other will allow early identification of potential foci of infection among office workers, or even the population of districts and cities. As an example, we can again mention the Safezone SafeTag device from KINEXON. This device is not only a distance meter, but also can be worn as a bracelet or simply attached to clothing. SafeTag very accurately and quickly determines the distance between people and in real time notifies the owner of any violations. Moreover, with the help of the extension of the settings of the software of this device, you can track the chains of infection, adjust the duration of contacts, if necessary, and even announce an alarm, warning employees about the danger.

In Belgium, the port of Antwerp used bracelets to maintain social minimum distance in the workplace [3]. The Dutch company Rombit [4] creates these devices. They alert the wearer of social distancing violations, and can track contacts, which can help identify chains of COVID-19 infection. These bracelets have built-in Bluetooth and the latest generation of ultra-wideband (UWB technology), which makes them premium devices. All information collected from the bracelets are stored on Rombit servers for further information processing. However, wearing such bracelets, sensors and using applications is not compulsory, and everyone has the right to decide for himself how to dispose of his personal information. However, if at least 60% of the population used these capabilities of the IoT, it would be much easier to control and suppress the spread of COVID-19.

2.4 EARLY DIAGNOSIS OF INFECTION

Not only the observance of social distance can effectively resist the virus, but also the early diagnosis of infection, which makes it possible to ensure that the disease does not spread by isolating the patient in time. The incubation period of the disease is long and imperceptible, but during this time, a person is a carrier of COVID-19 and can infect others, whether they are close friends, family members, coworkers, or even just passersby. Using the capabilities of the IoT, it is possible to track the chain of infections and isolate possible carriers in time.

Thus, in Hong Kong, special bracelets were introduced, connected to smartphones, which are issued to people who arrived from different provinces of mainland China [5]. These bracelets were connected to the patient's smartphone and monitored the movements of the alleged infected, preventing them from violating the established preventive quarantine and hiding from observation. When there is a violation of home quarantine, a message will be sent to the authorities (police) and the violator will be sent to a quarantine center. Furthermore, in China, a smart helmet and glasses are used to determine the temperature of people in public places. They replaced a portable thermometer, which was inferior in accuracy and speed. Thermal imaging cameras measure the temperature of objects and send the location of the infected using the built-in GPS, along with his photograph, to the officer's smartphone patrolling the area.

The use of the IoT has significantly helped improve methods for recognizing a dangerous virus in the early stages, which in turn reduces the risk of new infections. Along with thermal imaging cameras and goggles, gadgets that are attached to the body and connected to a smartphone through special applications have become widespread. One of these gadgets is the iFeel-You smart bracelet from IIT – Istituto Italiano di Tecnologia, which sends a warning to the owner's smartphone if its temperature rises above 37.5°C. The bracelet can also measure the distance to another bracelet owner using radio signals, and in case of violation of the distance between the owners, report this by means of vibration [6]. Another company called BioIntelliSense has invented a coin-sized sensor. The BioSticker [7] attaches to the upper left side of the chest under the clothing and continuously collects biometric data such as respiratory rate, body temperature, and heart rate, which allows you to quickly calculate the first signs of respiratory infections. All changes in a person's state are sent to his personal mail, where he can view the entire "history" of his body's life. BioIntelliSense has also released a more advanced and sharpened specifically for the COVID-19 line of sensors called BioButton [8]. Moreover, the 2021 MedTech Breakthrough Awards voted BioButton as The Best New Monitoring Solution.

In addition, a new coronavirus study has begun in Liechtenstein [9]. The Principality is funding the COVI-GAPP medical study, which is being carried out by the Swiss diagnostic company Dr. Risch Medical Laboratory. They intend to use the Ava brand's biometric bracelets to gather information on a person's health. A person puts on the bracelet at night and the device reads movements, body temperature, blood circulation, respiration, and pulse. Then, the algorithm calculates whether the owner of the device has early symptoms of COVID-19 (shortness of breath, cough, or fever). More than two thousand people (approximately 5% of Liechtenstein's population) took part in this study [10].

2.5 REMOTE PATIENT MONITORING

With overcrowded hospitals and the temporary closure of other health facilities, there is a risk that a lot of patients will not receive the care they need. When health workers began to fall ill, the government introduced safety requirements related to isolation, and it became clear that to save lives, it was necessary to implement remote health monitoring systems. The way out of this difficult situation was the use of the IoT, which in turn led to the creation of telemedicine.

Telemedicine is the remote provision of medical services (e.g., patient monitoring and consultation) and the interaction of medical workers with each other using telecommunication technologies. With the telemedicine development, people save time and effort, and it is also safer during a raging pandemic, because they can communicate with a specialist online. Neither patients nor doctors themselves are at risk of infection by avoiding physical contact and long hospital queues. Moreover, the use of such technologies creates the possibility to observe in real time the condition of the patients who require regular examinations and remind them of taking medications. Online consultations with several doctors (Figure 2.3), analysis of analyzes, and timely medical intervention in case of an urgent situation are possible. In addition, doctors can communicate not only with patients, but also with each other: discussing test results or patient histories, online broadcasting of surgery, consultations with foreign doctors. Mobile applications, video conferencing, virtual and augmented reality technologies, wearable gadgets that can send data to the doctor about temperature, pressure, pulse, and physical activity of the patient – all of them are used in telemedicine to work more efficiently regardless of place and time [11].

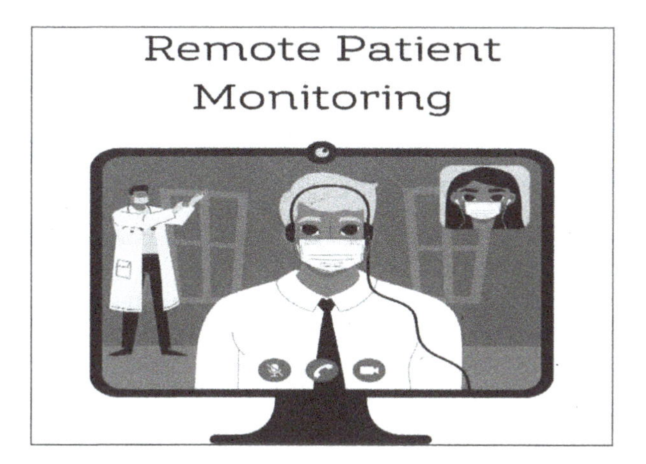

Figure 2.3 Video call with doctors.

The Internet of Medical Things (IoMT) use case has become a lifeline for healthcare facilities during the exacerbation of the pandemic. In addition, to all this, the use of remote monitoring has reduced the risk of contracting the virus for the most vulnerable segment of the population – the elderly. All kinds of sensors and portable devices attached not only to the body, but also to the patient's bed made it possible to gather all the necessary information on the present health status of patients, helped to prevent deterioration in health, and reduced physical contact between the infected and the medical staff to a minimum. For example, in China, the Shanghai Public Health Clinical Center and seven other hospitals have used body-connected thermometers to constantly monitor the temperature of their patients. This monitoring helped to reduce the risk of cross contamination of nursing staff from patients and made it possible to increase the effectiveness and safety, visiting the patient only when it is necessary or in unexpected critical situations. VivaLNK body temperature sensors worked together with the Cassia IoT Access controller. All information about changes in the patient's body temperature was transmitted to the nurse's post in real time, which eliminated unnecessary physical contact [12].

HD Medical has developed an all-in-one remote patient monitoring technology. This gadget is called HealthyU [13] and is a small device that includes an intelligent stethoscope, a seven-lead ECG, it can also measure heart rate, blood oxygen saturation, temperature, blood pressure, and respiratory rate. Moreover, this device has an audio jack using which you can broadcast the sounds received by a stethoscope through a video or audio call. HealthyU is so easy to use that even our grandparents, who do not understand anything about modern technologies, can handle its operation: all you need to do is touch the special connectors with four fingers. All read data are displayed on the owner's smartphone using a connected application, and they are sent to storage on a cloud platform, through which the patient's doctor can easily view them. Using this device will help take remote patient monitoring to a completely new level [10].

2.6 QUARANTINE CONTROL

Quarantine control is another great IoT use case for controlling the spread of COVID-19 (Figure 2.4).

This was mainly applied to people who had just returned from a trip abroad, or to those who were believed to be infected. Various GPS trackers were used that were built into mobile phones or other IoT devices, which allowed the authorities to monitor movement and quarantine compliance by patients or suspected infected. Smart wearable tracking devices have been used by South Korea, Europe, the Middle East, and Singapore to determine if travelers have recently violated quarantine or not.

Figure 2.4 A person working from home because of quarantine.

Thus, during periods of strong spread of coronavirus in many provinces of China, a color-coded system for citizens was introduced, since the Chinese government did not see the need to establish a complete and widespread lockdown. Therefore, Chinese citizens are registered in the Alipay Health Code application through which they are assigned a specific color code (green, yellow, or red). This code is generated based on a questionnaire that each user fills in upon registration. Having received a green code, a person has the ability to move freely, with a red one, a two-week quarantine should be required, and with a yellow code, it will be necessary to refrain from any movement for one week. Moreover, data about color codes of citizens are available to the country's police, which helps to quickly find violators [14].

In South Korea, information about the movements of people in quarantine comes from GPS data or bank card transactions. In Turkey, the control of compliance with quarantine is monitored through the data of mobile networks: in case of a violation of quarantine, a warning is sent to a person, they can also receive a call reminding them of the need to observe isolation and to protect those who are infected and their surroundings. In Poland, the home quarantine application is used, which is applied to people who are required to sit in quarantine for two weeks after arriving from abroad. The user must send his photo and geolocation data within 20 minutes after receiving the request, but if the request remains unanswered during the specified period of time, the police receive a warning about the violation of quarantine.

In Bahrain, people who are medically imposed on mandatory quarantine must download the government-approved contact tracing application

BeAware Bahrain. Internet and Bluetooth must always be turned on, and people must indicate their quarantine location. Moreover, people need to wear special bracelets with a built-in GPS function, which are subsequently connected to the application. So the government can easily control the movement of people under quarantine [15].

2.7 IOT APPLICATIONS

The IoT system has a close connection to mobile applications through which users have the ability to manipulate information and to control various gadgets and sensors. The application market is insanely huge and continues to grow all the time. Their functionality can surprise even the most inveterate cynic, not to mention the common person. During a pandemic and a complete lockdown, special attention has begun to be paid to applications that can help in the fight against coronavirus. For example, applications which can monitor compliance with quarantine and mask mode by tracing contacts with infected or possibly infected people. Moreover, the development of such applications was taken up by the governments of the countries themselves, in order to protect their citizens.

Table 2.1 presents a list of mobile applications that actively help in the fight against the spread of coronavirus in different countries all around the world. They are actively used and anyone can download it. Also, some of the applications can work together with special bracelets, mainly those applications that monitor compliance with quarantine or can analyze biometric data. In addition, there are purely informative applications that are used to inform the people about the current situation with the coronavirus, the number of infected and other important information. Online consultations with doctors, self-diagnosis through special questionnaires, making appointments with doctors through online offices, viewing a personal medical record through your phone – this is just a small part of what can be done using the applications listed in Table 2.1.

Table 2.1 The list of applications

No.	*Application name*	*Country*	*Function*
1.	CovidSafe	Australia	• Helps to identify if users have contacted people with coronavirus.
2.	Rakning C-19	Iceland	• Helps to identify if users have contacted people with coronavirus. • Builds a safe route.
3.	VírusRadar	Hungary	• Helps to identify if users have contacted people with coronavirus.

Table 2.1 Cont.

No.	Application name	Country	Function
4.	CoronApp	Colombia	• Helps to identify if users have contacted people with coronavirus. • Self-diagnostic. • Gives recent information about the spread of COVID-19.
5.	eRouška	Czech	• Helps to identify if users have contacted people with coronavirus.
6.	CoronaCheck	India	• Quick self-diagnostic.
7.	InfoCU	Cuba	• Gives information about COVID-19. • Gives recommendations and instructions to prevent the risk of infection.
8.	Corona Map	Saudi Arabia	• Self-diagnostic. • Gives recent information about the spread of COVID-19.
9.	Alipay Health Code	China	• Helps to identify if users have contacted people with coronavirus. • Quarantine enforcement control.
10.	NCOVI	Vietnam	• Gives information about COVID-19. • Gives recommendations and instructions to prevent the risk of infection.
11.	SpeetarHealth	Libya	• Self-diagnostic. • Free online communication and consultation with doctors.
12.	Ito	Germany	• Helps to identify if users have contacted people with coronavirus.
13.	Estamos ON	Portugal	• Gives recent information about the spread of COVID-19.
14.	ViruSafe	Bulgaria	• Self-diagnostic. • Free online communication and consultation with doctors.
15.	DOCANDU Covid Checker	Greece	• Free online communication and consultation with doctors. • Open access to a complete medical record.
16.	Self-Safety	Uzbekistan	• Helps to identify if users have contacted people with coronavirus.
17.	"HaMagen"	Israel	• Helps to identify if users have contacted people with coronavirus.
18.	Hayat Eve Sıar	Turkey	• Gives information about COVID-19. • Gives recommendations and instructions to prevent the risk of infection. • Quarantine enforcement control.
19.	SOS-Covid	Ecuador	• Quarantine enforcement control. • Diagnostic. • Free online communication and consultation with doctors.
20.	TraceTogether	Singapore	• Tracks with whom the user contacted and stores this data for 21 days.

2.8 FUTURE OF THE INTERNET OF MEDICAL THINGS

Even some 20 years ago, no one could have imagined that modern digital technologies would change our daily life in such a way and become an integral part of it. Flying drones, multifunctional surveillance cameras, all kinds of sensors that read our biometric data, robotic vacuum cleaners, smart speakers, smart homes, and many more different devices surround us every day. The huge networks created by the IoT are constantly working, generating a huge amount of incoming information, while never ceasing to improve. A huge impetus to the even greater development of the IoT was the global coronavirus pandemic: work, study, shopping – everything began to work with the help of the Internet. However, one of the fastest growing areas of IoT development has become healthcare (the development of IoMT and telemedicine).

The development of 5G Internet technology with a speed of more than 100 mb/s in the near future will allow combining various technological devices and expanding their coverage areas, since at the moment there are zones remote from large settlements without Internet access. Such improvements will allow people in remote regions to access advanced medical services. Medical devices connected using high-speed Internet will be able to collect and transmit data about patients over a distance in seconds, and the development of remote diagnostics, video conferencing, remote surgery, and telemedicine will make it possible not only to overcome the geographical gap and the level of medical services, but also to increase the productivity of medical care in emergency situations and reduce their cost to patients.

In addition, while increasing the capabilities of medical devices using Big Data technology, it can increase the accuracy of diagnoses and reduce the number of medical errors that can be caused by the human factor. Wearable measuring devices have already become a part of the lives of many patients around the world. They, for example, help control blood sugar levels; remind you to take medications on time; read the pulse and blood pressure; automatically measure the body temperature and notify the wearer if it rises strongly (which is extremely effective during a pandemic for the early diagnosis of COVID-19).

The possibilities of telemedicine and the use of robotics will make it possible in the future, for example, to carry out surgical operations for patients who are in hard-to-reach regions or in conflict zones. Today, the use of robots during operations involves the obligatory participation of doctors in this process. Therefore, scientists are working to create robots that can minimize human participation in operations. Moreover, in addition to helping with operations, robots help in the diagnosis and therapy of patients. Transplantation also involved robotics; bioprinters are widely used in it. They are already able to print scaffolds of tissues, organs and hyperelastic bones, placenta models using a liquid nutrient substrate with different types of living cells, gels, fibers, polymers, ceramics, metals, and other materials.

Their development will help in the future to grow new full-fledged organs from the cells of the patient himself, which will drastically reduce the risk of rejection of the transplanted organs by the body [15].

The development of the IoMT in the future will help bring medicine to a whole new level. It will become both more affordable and quality at the same time. There will be constant access to medical data of patients and the ability to quickly diagnose, consult with doctors right from home, remote monitoring of patients' condition, and early diagnosis of diseases.

2.9 CONCLUSION

The use of modern digital technologies of the IoT in the fight against coronavirus has helped humanity to avoid even greater losses among the population. The IoT is building a vast healthcare network to fight the coronavirus. All necessary data is stored online and, if necessary, can be easily accessed by both the patients themselves and their attending physicians. Infected patients or those who are at risk and at the same time closed in home quarantine can freely communicate with their attending doctors through online conferences, see the results of their analyzes through personal offices, and receive the necessary treatment instructions without leaving home. In addition, online interaction with the treating doctor helps people living in remote areas to receive highly qualified medical care. With the help of special gadgets and applications, you can maintain social distance and prevent contact with infected people; receive the most recent news about the epidemiological situation in the country; as well as diagnose earlier infection with a virus using gadgets that read biometric data.

REFERENCES

[1] "Safezone: A New Device That Can Help Slow the Spread of Covid-19", 20 October 2020, Newswire, www.newswire.com/news/safezone-a-new-device-that-can-help-slow-the-spread-of-covid-19-21239327. Accessed: 21 August 2021.
[2] Bunina, V., "Contact Trackers and Skin Sensors: How Gadgets Help Fight COVID-19", 16 November 2020, Gazeta.ru, www.gazeta.ru/tech/2020/11/16/13362691/gadgets_covid.shtml?updated. Accessed: 17 August 2021.
[3] Meisenzahl, M., "Dock Workers in Belgium Are Wearing Monitoring Bracelets that Enforce Social Distancing", 9 May 2020, Insider, www.businessinsider.com/workers-in-belgium-use-social-distancing-wearable-tech-2020-5. Accessed: 17 August 2021. https://rombit.com/covid-solutions/
[4] Doffman, Z., "Coronavirus Police Surveillance Tags Are Now Here: Hong Kong First to Deploy", 17 March 2020, Forbes, www.forbes.com/sites/zakdoffman/2020/03/17/alarming-coronavirus-surveillance-bracelets-now-in-peoples-homes-heres-what-they-do/?sh=6de97cce4533. Accessed: 18 August 2021.

[5] Siddiqui, S., Shakir, M.Z., Khan, A.A., Dey, I. (2021) "Internet of Things (IoT) Enabled Architecture for Social Distancing During Pandemic". *Front. Comms. Net* 2:614166. doi:10.3389/frcmn.2021.614166

[6] Whittaker, J., "Body Sensor Aims to Provide Early Warning System for COVID-19", 22 September 2020, Cayman Compass, www.caymancompass.com/2020/09/22/test-driving-the-biosticker/. Accessed: 21 August 2021.

[7] https://biointellisense.com/biobutton

[8] Martens, S., "Liechtenstein Relies on Biometric Wristbands in the Fight Against Corona", 20 April 2020, Netzwoche, www.netzwoche.ch/news/2020-04-20/liechtenstein-setzt-auf-biometrische-armbaender-im-kampf-gegen-corona. Accessed: 21 August 2021.

[9] Pennic, F., "Shanghai Public Health Clinical Center Uses Wearable Sensors to Combat Spread of Coronavirus in China", 31 January 2020, Hit Consultant, https://hitconsultant.net/2020/01/31/shanghai-public-health-clinical-wearable-sensors-coronavirus-in-china/. Accessed: 23 August 2021.

[10] http://deslab.uk/2020/04/29/novyj-smart-braslet-ifeel-you-sposoben-izmeryat-temperaturu-tela-i-obespechivat-socialnoe-distancirovanie.html

[11] www.businesswire.com/news/home/20210111005444/en/HD-Medical-at-CES-Unveils-HealthyU%E2%84%A2-the-World%E2%80%99s-First-Intelligent-All-in-one-Remote-Patient-Monitor-for-Telehealth-and-Wellness

[12] Mozur, P., Zhong, R., Krolik, A., "China Coronavirus Surveillance", 1 March 2020 (updated: 26 July 2021), New York Times, www.nytimes.com/2020/03/01/business/china-coronavirus-surveillance.html. Accessed: 23 August 2021.

[13] www.intel.ru/content/www/ru/ru/healthcare-it/telemedicine.html

[14] 'BeAware Bahrain' App: https://healthalert.gov.bh/en/category/beaware-bahrain-app

[15] Robots in medicine: applications and opportunities, 2 August 2019 https://top3dshop.ru/blog/the-latest-medical-robots.html. Accessed: 28 August 2021.

Chapter 3

Internet of Things (IoT)-based smart farming system

A broad study of emerging technologies

Kamlesh Gautam, Arpit Kumar Sharma, Amita Nandal, Arvind Dhaka, Gautam Seervi, and Shivendra Singh

CONTENTS

DOI: 10.1201/9781003230236-3

3.1 INTRODUCTION

To upgrade the rural produce with fever resource and work endeavors, major advancements have been made all along with the human set of wisdom. The immense population rate nevermore let interest and supply match throughout each of these times. By 2050, 9.8 billion population will be associated virtually, which makes 25% of the total population among agricultural countries. On the opposite side, the criterion of urbanization is determined to advance at a speedup, with about 71% of the world's population partner manager organizing the audit of the article and endorsing it for distribution was Kun Mean Houston anticipated to be metropolitan until 2050 (at present 48%) [1]. Besides, pay extent will be products of what they are in present, which will operate the daily bread request advance, especially in agricultural countries. Subsequently, these nations will be more attentive about their consuming routine and food grade; thus, shopper inclinations can shift from grains to greens and, later, to the flesh.

To look out this sizeable, more metropolitan, and more spendthrift populace, feed creation must be twofold by the 2050s. Particularly, the current figure of 2.2 billion tons of annual oat creation shall contact roughly 3.2 billion tons, and the annual meat creation shall increment near to more than 200. Because Internet of Things (IoT) is most useful for agriculture regarding monitoring, controlling, prediction, and logistics as described in detailed structure using Figure 3.1.

3.2 SIGNIFICANT APPLICATIONS

By carrying out the current recognition and IoT growth in Agro business practice, each segment of conventional farming techniques can be generally altered. Currently, a compatible mix of remote detectors and IoT in acute Agro business can uplift horticulture to the extent that was earlier inconceivable. By succeeding the action of brilliant Agro business, IoT can help with enhancing the adjustment of diverse conventional farming issues, identical to dry season return, and appropriateness, irrigation system, yield enhancement, and inflammation control. We will consider the reports on the importance of important applications, remote detectors, and administrations being utilized for savvy horticulture operations. This book chapter discusses major occasions in which cutting-edge innovations are assisting at different levels to improve general proficiency in agricultural sector [2].

3.2.1 Soil sampling and mapping

Soil is referred as "stomach"' for floras, and inspecting is an initial move of evaluation to acquire farm-specific data, that is later used for resolving distinct primary alternatives at various levels [3]. The basic aim of top-soil

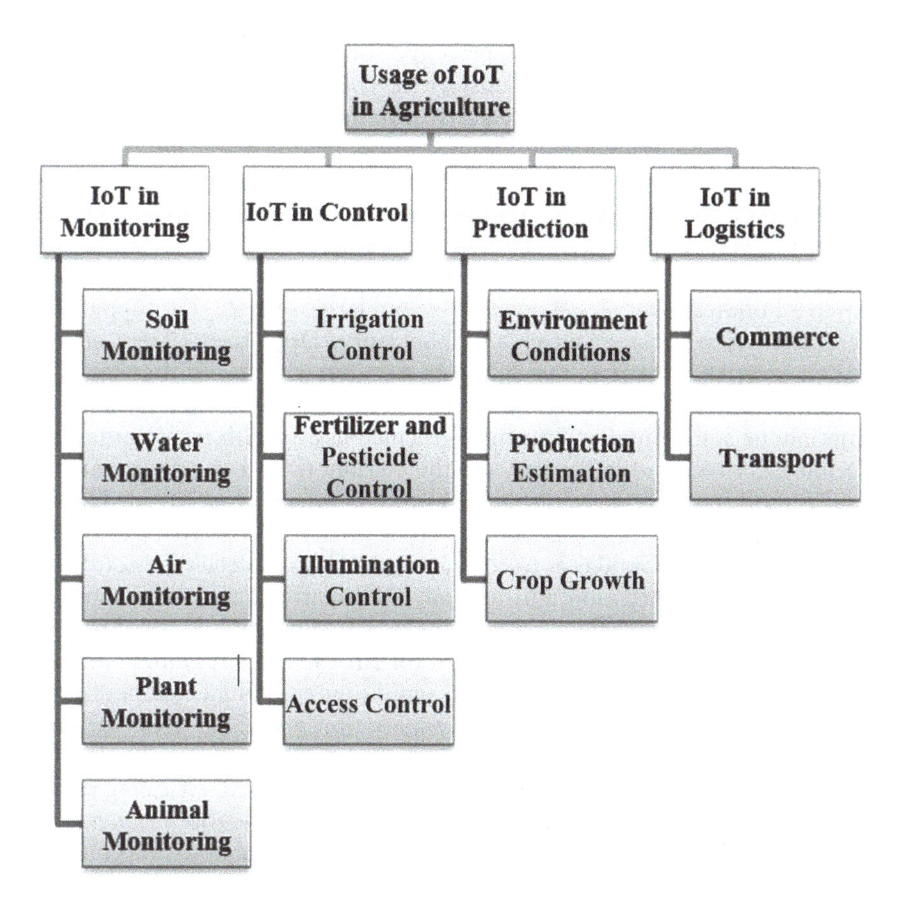

Figure 3.1 Hierarchical structure of usage of IoT in agriculture.

inspection is to find the supplement level of the land so that necessary actions could be taken for further improvements. The factors which are primary to examine the dirt supplement extent assimilate trimming records, geography, compost application, soil type, irrigation system level, and so on. These elements grant understanding in regard to the physical, composite, and natural situations with dirt to identify the limiting elements to a level so that the yield can be controlled likewise. Soil planning compels the method for settling diverse harvest variety in a specific farm to more willingly synchronize with soil attributes appropriately, likely to seed appropriateness, time to sow, and surprisingly the growing acuity, as some are intense habitual and remaining less. Apart from this, forging numerous yields simultaneously can also bring out more intelligent usage of farming, essentially utilizing assets. At present, makers are giving a wide scope of tool compartments and sensors that can help ranchers to follow the dirt quality and, in light of this information, prescribe solutions to keep away from its corruption. These frameworks consider the checking of soil properties, like surface, assimilation rate, and water-absorbing limit that eventually aid to limit salinization, fermentation, disintegration, densification, and contamination (keeping away from unreasonable utilization of compost). Laboratory in the box, a dirt trial tool compartment created by Agro cautions, is viewed as the total lab in itself, dependent on its provided administrations. Utilizing this way, any rancher, besides having any laboratory experience, can dissect up to a hundred examples every day (by and large, over 22,000 supplement tests a year) without going to any laboratory. Dry seasons are a significant matter that restricts the efficiency of harvest yield. To manage such issues, particularly for rustic regions, far-off detecting is being utilized to get continuous soil dampness information which assists with investigating the farming dry spell in far districts. For this reason, the ocean salinity satellite and topsoil moisture were dispatched in 2009 which gives worldwide soil dampness maps. In this exertion, they come after various ways to deal with getting dirt water boundaries to contrast and soil water deficit index (SWDI) obtained from in-situation information. The creators utilized the common goal imaging spectroradiometer detector to plan different soil useful properties to assess the soil corruption hazard for Africa. The dirt guides and land study information, which enclosed all significant environment zones on the mainland, were utilized to foster the expectation prototype. Detectors and sight-based advancement are useful to choose distance and profundity for planting seeds productively. As in detector and sight-based self-ruling robots known as Agro Bot are produced for planting seeds. The robot can execute on either agrarian ground on which mindfulness of robot arrangement is determined by the worldwide and nearby guides produced using GPS while on-board sight framework is matched with the PC. Progressing further, different non-consent detecting strategies are planned to decide the seed stream speed where the detectors are outfitted with a light-emitting diode; comprised of IR, noticeable light, and laser-LED just like a component as an emission beneficiary. The yield energy shifts reliant upon the improvement of the seeds through the sensor and band of light shafts, and the falling of shades on segments of the gatherer. The sign data, connected to the passing seeds, are utilized by the gauge to seed the stream rate.

3.2.2 Water system

Nearly 97% of Earth's water is salinely acquired by seas and oceans, and leftover 3% is new water, further, two-third is frozen in the types of glacial masses and polar ice covers. Just 0.5% of dissolved new water is over the land or noticeable all around, as the rest is underground. To put it plainly, humankind depends on this 0.5% to satisfy every one of its prerequisites and to keep up the biological system, as ample new water should be retained in waterways, and another comparable repository to support it. The referencing metrics of new water system exclusively utilizes 70% of horticulture business. Talking about the numbers of water extremity throughout the planet, simultaneously its enlarging appeal in agribusiness and various distinct enterprises, it needs to be delivered to areas like where it is needed, in particular, in essential amounts. For this cause, enlightened attentiveness has been carried out to save the present under-stress water resources by handling more constructive water scheme frameworks. Distinct managed water system methods, like as the dribble method and sprinkler method, are being upraised to control the water wastage problems, which were similarly found in conventional methods like the flood method and wrinkle method. Both yield amount and quality are gravely impacted while facing water lack, as sporadic water methods, even ample, cause declined top soil supplements and encouraged distinctive bacterium diseases. It's a basic undertaking to specifically assess water interest of yields, where components such as harvest type, water system methods, crop needs, topsoil type, rain, and soil moisture upkeep are combined. Talking about this harsh reality, an accurate land and air moisture management framework handling the remote detector requires water and urge better yield benefits. The current circumstance of water system techniques is relied upon to be altered by receiving arising IoT advancements. A huge expansion in farm productivity is normal with the utilization of IoT-controlled procedures, for example, crop water pressure list-based water system, etc. For this, accomplishing farm shelter at the various course and air temperature are required for the estimation of factors. A remote sensor and detector-based observing framework, where all land sensors are associated, communicate with each other and gather referenced estimations to compare the programming operations in order to utilize dissecting homestead information. This as well as data from different sources, including climate information and satellite imaging, is applied to controlled models for water required appraisal; lastly, explicit water system list esteem is created for each site as in the form of Big Data [4]. An unmistakable model is variable rate irrigation streamlining by crop metrics that works as per geology or topsoil inconstancy, at last reform the water utilization productivity.

3.2.3 Manure

Manure is a synthetic substance that can give considerable supplements for the fruitfulness, nourishment, and development of plants. Plants predominantly need three main macronutrients: phosphorus for blossoms, root, and organic product advancement; potassium for water and stem development; nitrogen for leaf development. Any type of supplements deficiency or using them unsuitably can be hurtful for the plant. Talking more significantly, unnecessary usage of compost leads to monetary inconvenience and makes unsafe effects on climate and dirt by fatigue the dirt quality, maltreating groundwater, and leading to worldwide environmental changes. In general, crops consume not absolutely a big portion of nitrogen used as manure, while leftover either discharged to air or vanished as drainage. Uneven usage of manure results in an unevenness in soil supplement levels and the worldwide environment as nearly 80% of the world's deforestation has occurred due to agrarian practices only. Treatment under shrewd agribusiness plays to unconditionally appraise the essential part of supplements, at last limit their adverse results on climate. Studies feel the necessity for site-explicit soil supplement level evaluation reliant on distinct elements, such as soil type, richness type, and use rate, soil ingestion capacity, item yield, climate condition, crop type, and many more. The evaluation of topsoil supplement level is quite expensive as it needed the testing of soil at each region. New IoT-based preparing approaches aid to appraise the structural examples of additive prerequisites with greater attention and minimum work necessities. For precedent, the normalized difference vegetation index (NDVI) uses airborne/satellite images to examine crop nutrient status. Sometimes many satellite images work or find the entity like medical image disease [5]. Essentially, NDVI relies on the reaction of observable and close IR rays from flora and is used to measure the harvest health, thickness, and energy, further computed to gauge the dirt supplement grade. Such compel implementation can overall reform the compost proficiency, all the while reducing the results to the climate. Several new legitimized revolutions, alike variable rate technology, geo planning, self-governing vehicles, and GPS precision, are firmly supporting IoT-based shrewd treatment [6]. Moreover, fertigation, accuracy dealing, and chemigation are more rewards of IoT. In these procedures, water-solvent issues, like compost, soil corrections, and pesticides, could be useful through the irrigation system framework. Albeit, such approaches are not newfangled to farming and have been functional over most current 30 years, their precise usage with real outcomes has been seen uniquely along with IoT reconciliation. Because of continuing results, fertigation is measured as the finest administration preparation to advance the viability of other horticulture kinds of stuff; in particular, it inclines to be coordinated with IoT-based savvy farming framework flawlessly and some time working style as like medical image resorption or planning to how to handle IoT-based system or sometimes find the loop pole research concern type of networks like deep learning network and techniques [7].

3.2.4 Yield observation, anticipating, and accumulation

Yield observing is an instrument castoff to examine different perspectives relating to farming yield, similar to grain mass stream, dampness content, and gathered grain amount. It serves to precisely evaluate by reporting the harvest produce and dampness grades to appraise, how fine the yield function, and what needs to be done straight away. Yield observing is viewed as a fundamental piece of exactness cultivating at the hour of gather as well as even earlier that, as checking the yield quality assumes a significant part. Yield quality relies upon numerous elements, for example, adequate fertilization with great quality dust particularly when anticipating seed yields under changing ecological conditions as of now, when we are managing more open business sectors, purchasers throughout the planet become more specific about natural product quality; thus, viable creation relies upon the correct natural product scope to the perfect marketplace at the perfect instance. Harvest gauging is craftsmanship toward anticipating the produce and creation before reap happens. These anticipating aids the rancher meant for not-so-distant forthcoming arranging and dynamic. Moreover, breaking down the product quality and its development is an extra basic feature that empowers assurance of the ideal period for reaping. This observation counterbalances different advancement phases and utilizes natural product circumstances like their size, tone, and so on. IoT-based shrewd cultivation is toward making the farms talk not just assists with augmenting the harvest quality and creation yet, in addition, gives a chance to change the administration methodology. Although the gathering is the last phase of this interaction, legitimate booking can have an unmistakable effect. To acquire genuine advantages from crops, ranchers need to know when these yields are prepared to collect. This technique is used in where creators utilized Sentinel-1A Interferometric pictures for planning the rice crop produce and power in Myanmar. As already we have referenced before in this part, natural product extent consistently assumes basic part to assess development, settling on choices for gathering and focusing on the correct market, for this reason, shading (RGB) profundity pictures are utilized to follow the diverse natural product conditions in mango ranches.

Likewise, different visual sensors are utilized to screen the contracting of papayas, particularly through drying situations. To make this point more clear, the IoT-based Farming Monitoring Model is explained in Figure 3.2, which uses HVAC modules, sensors, and actuators with Control Unit and Database Server including Weather Forecast Provider.

3.3 ADVANCED AGRICULTURAL PRACTICES

Receiving the clever techniques to upgrade the quality and amount of food is not another thing, as people have been doing this for quite a long time. At first, we attempted to improve the harvest creation by zeroing in on seed assortment, composts, and pesticides. Before long, it was understood that these ordinary ways were not sufficient enough to fit this interesting hole; consequently, horticulture researchers have started considering different other options, as bioengineered food varieties, otherwise called hereditarily

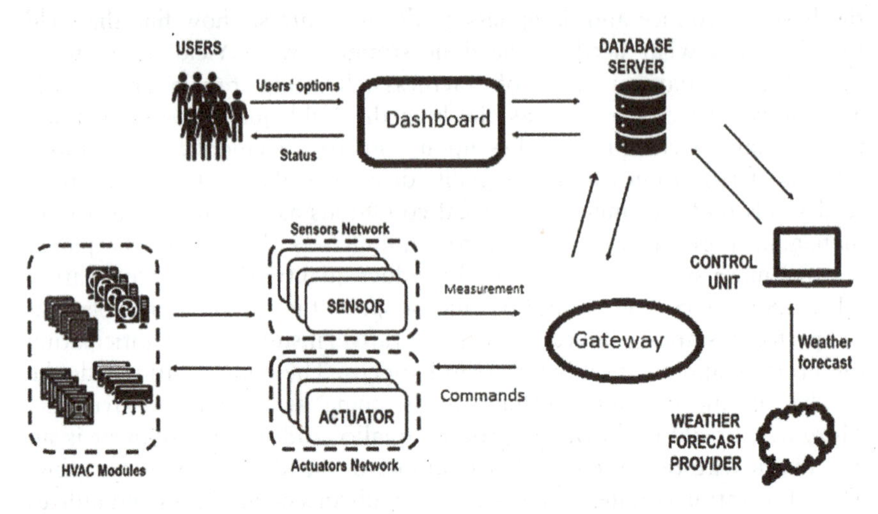

Figure 3.2 IOT-based farming monitoring model.

adjusted or hereditarily designed food varieties, are food sources delivered by bringing changes into their DNA utilizing the strategies for hereditary designing. Be that as it may, a few investigations feature their genuine consequences for human well-being, including barrenness, interruption in safe framework, sped up maturing, defective insulin guidelines, and so forth. Every one of these and numerous other comparable advances didn't get a lot of notoriety and acknowledgment in the public arena since individuals incline toward bio and natural food. In this respect, enormous exploration has been directed for quite a long time in which sensors and IoT-based innovations are assisting with further developing traditional agribusiness cycles to upgrade yield creation without, or with least, impact on its inventiveness. For this reason, new refined and more controlled conditions are projected to handle the previously mentioned issues. The significance and inclusion of new advances are more basic, as we are pushing toward more refined and metropolitan cultivation. Indeed, it would not be inaccurate for one to say that the achievement of these high-level practices is in question without utilizing sensor-based innovations.

3.3.1 Green house farming

Nursery cultivating is viewed as the most seasoned technique for keen cultivating. Albeit, developing plants in a controlled climate isn't new as found since Roman occasions yet it acquired fame in the nineteenth century where biggest nurseries were implicit France, Netherlands, and Italy. Further, the training was sped up during the twentieth century and profoundly advanced in nations that confronting brutal climate conditions. Harvests developed inside are exceptionally less influenced by climate; above all, they are not restricted to getting light just during the daytime. Therefore, the yields that customarily must be developed under appropriate conditions or in specific pieces of the world are presently being developed whenever and anyplace. This was the real time wherein sensors and specialized gadgets began to help different farming applications truly. The achievement and creation of different yields under such a controlled climate rely upon numerous variables, similar to exactness of checking boundaries, design of shed, covering material to control wind impacts, ventilation framework, choice emotionally supportive network, and so on. A definite investigation is given in where this load of variables, their effects, and how remote sensors can help for this are thought of. Exact observing of climate boundaries is the most basic errand in present-day nurseries, where a few estimation points of different boundaries are needed to control and guarantee the nearby environment. An IoT-based model is proposed to screen the nurseries where mica hubs are utilized to gauge within boundaries like stickiness, temperature, light, and pressing factor.

3.3.2 Vertical farming

The world necessities more farmable grounds to satisfy expanded food requests; however, the truth is that 33% of arable land was lost during the most recent 40 years because of disintegration and contamination. Sadly, current agrarian practices dependent on modern cultivating are harming the dirt quality far quicker than nature can modify it. In general, it is assessed that disintegration rates from developed fields are ten to multiple times more prominent than the dirt arrangement rates. Considering the decrease of arable land issues, it very well may be a calamity for food creation soon with current farming practices. Further, as we referenced, 70% of new water is just utilized for agribusiness reason, which can expand the weight on existing restricted water supplies. Vertical farming (VF) is a response to address the difficulties of land and water deficiencies. VF, as metropolitan farming, offers a chance to stack the plants in a more controlled climate coming about in, above all, huge decrease in asset utilization. By following this technique, we can expand the creation on various occasions, as just a small part of ground surface is required (contingent upon the quantity of stacks) when contrasted with customary horticulture rehearses. Not just for ground surface, this framework is exceptionally proficient as far as different assets, also. The figures are exceptionally promising, as it is creating 10,000 heads of lettuce each day (twofold the creation when contrasted and customary strategies) and are, in particular, burning-through 40% less energy and up to close to 100% diminished water utilization contrasted with open airfields. Air cultivates, an innovator in VF, developing horticultural items with up to multiple times more significant returns while using 95% less water at Newark. Under this cultivating strategy, numerous boundaries are significant; however, CO_2 estimations are generally basic; thus, nondispersive infrared (IR) CO_2 sensors assume a basic part to track and control the conditions in vertical ranches. Boxed Giscard, created by Edinburgh Sensors, is particularly planned by thinking about such a climate, which utilizes a pseudo double pillar NDIR estimation framework to upgrade the security and decrease optical intricacy. Human hands are not needed to contact the yields at any stage when following the IoT-associated vertical homestead; this is the case made by Mint Controls engineers who offer a wide scope of arrangements, similar to squander compartments and sensors and their incorporation for different VF applications.

3.3.3 Hydroponic

To improve the advantages of nursery cultivating, horticulture specialists pushed ahead another progression and gave the possibility of tank farming, a subset of hydroculture in which plants are developed without soil. Tank farming depends on a water system framework in which adjusted supplements are disintegrated in water and yield establishes stay in that arrangement; now and again, roots can be upheld by a medium like a perlite or rock. When consolidating tank farming with VF, a homestead of 100 square meters can create the yield identical to one section of land of conventional ranch, in particular up to 95% less water and composts usage and without pesticides/herbicides. At present, accessible frameworks and sensors are not just used to screen the scope of boundaries and take readings at predefined stretches at the same time, likewise, the estimations are put away so that they can be utilized to examine and demonstrative reason later on. Under this application, the exactness of supplement estimations is vital, thusly, an exceptionally dependable remote control framework for tomato aqua-farming is proposed in which they zeroed in on different correspondence norms that are least affected by plant quality and their development. The observing of arrangement substance and their exactness is generally basic under this strategy; for this reason, numerous frameworks are offered to check the presence of substance considering the plant requests. A remote sensor-based model is proposed to convey a turnkey answer for the tank farming development which offers ongoing estimations for soilless indoor developing. Further, a minimized sensor module is introduced which utilizes oscillator circuits to quantify the presence and groupings of different supplements and water levels.

3.3.4 Phenotyping

The recently examined shrewd strategies look more encouraging for the fate of agribusiness, as they are as of now being utilized to deliver diverse yield items under exact conditions. Other than these, a couple of cutting edge methods are under test to additional improve the yield capacities by controlling their impediments with the assistance of cutting edge detecting and correspondence advances. Among these strategies, the more unmistakable is phenotyping, which depends on arising crop designing, which connects plant genomics with its ecophysiology and agronomy. The advancement in subatomic and hereditary apparatuses for different harvest rearing was huge somewhat recently. Nonetheless, a quantitative examination of the harvest conduct, for example, grain weight, microorganism obstruction, and so forth, was restricted because of the absence of proficient procedures and innovations that we would now be able to appreciate. The experiments produce sufficient data for the reason that plant phenotyping can be exceptionally useful to explore the quantitative attributes, for example, those are answerable for its development, yield quality and amount, and opposition abilities to deal with different burdens. Additionally, the job of detecting innovations and picture-based phenotyping is featured in and portrays how these arrangements can assist with boosting the advancement for screening various bio energizers as well as their part in understanding the method of activities. Besides, an IoT-based phenotyping stage, Crop Quant, is intended to screen the harvest and significant characteristic estimations that can give the office to edit reproducing and computerized horticulture. Here, a programmed in-field control framework was created to handle the information produced by the stage. The AI characteristic investigations and calculations help to investigate the connection among the genotypes, aggregates, and climate.

3.4 SIGNIFICANT EQUIPMENT AND TECHNOLOGIES

Not quite similar to outdated farming, most of the assignments in current, enormous space agribusiness are being completed by substantial and sophisticated hardware, like collectors, farm trucks, and different robots that are supported with different detecting and various similar innovations. For precision agribusiness, whenever errands such as irrigation system, implanting, treating, and reaping are engaging in recreation, waged automobiles are furnished by GIS and GPS workplaces so they can effort decisively, area-specific, and self-ruling. Indeed, the possibility of the site-specific crop the boarding is absurdly deprived of counting the new tendency situation revolutions. The accomplishment of accuracy farming rests on the exactness of collected information, which is generally done in two diverse ways. The principal includes the application of multiuse exotic applications furnished with far-off sensing phases, like satellites, horticulture planes, inflatables, and UAVs; the additional is from distinct sorts of sensors—those that are usually taken for the unambiguous cause across dissimilar destinations of our benefit. The accumulated material is linked to the particular part of statistics by applying GPS devices so that area-explicit care could be provided subsequently. Agribusiness has changed during the most recent couple of a long time from little/medium cultivating activities to extremely commercial farming and industrialized. This growth grants the main partnerships to pact with horticulture like different ventures, for example, fabricating where approximations, statistics, and regulators are necessary to permit harmony among expenses and creation to support benefits. Appropriately, every part of farming that can be computerized, carefully arranged, and overseen will profit from IoT advances and arrangements. Because of this reality, endeavors are being engaged to offer more refined apparatuses like agrarian robots to show out a room of movements, such as planting, picking, irrigating, clearing, weakening, treating, splattering, persistent, and stirring. This transformation is being driven because of the headway of innovation, but at the same time as a result of elements like fear of mislaying minimal expense effort, in the specific necessity for improved and fewer exclusive sustenance. In light of this realism, through the time of 2020s, the worldwide keen cultivating marketplace is anticipated to increase at the development pace of 19.4% each year to contact $23.16 billion in 2023. Now it merits referencing that drones are creating and further predictable to produce the most elevated income among all horticultural robots used in brilliant cultivating. Perennial interest for larger harvest produce, expanded fuse of data, and correspondence innovation in cultivating and quick worldwide climatic variations are a portion of significant drivers coming about to like high marketplace development. Makers in the marketplace proposed an assortment of items and arrangements, for the most part, dependent on sensors and productive correspondence for a scope of uses; a couple has appeared in the key advances and hardwares that are right now accessible for this design are examined in after.

3.4.1 Modern tractor

As rustic work assets have begun to go under pressure because of the development of the yield business, farm trucks and other programmed large equipment began to enter the horticulture area. Where accessible, a normal size farm hauler can work multiple times quicker with altogether less costs than conventional ranch work. To satisfy the constantly expanding requests, agrarian-based gear makers, similar to Hello Tractors, Case IH, CNH, and John Deere, have begun to give improved arrangements zeroing in on producer's necessities. Through the headway of innovation, the vast majority of mentioned producers are presenting farm vehicles along with programmed operated and even cloud-computed figuring capacities. Such innovation isn't new-fangled, as driverless farm trucks are already present in the marketplace even already semi-self-governing vehicles. The primary benefits of driverless work vehicles are their capacity to try not to return to a similar region or line by lessening the cover even not exactly an inch. What's more, they can do extremely exact turns deprived of a driver's actual presence. This office bids well accuracy with diminished blunders, particularly when showering insect poison or focusing on weeds; those are for the most part inevitable when a man controls the apparatus. Even though, right now, no completely self-sufficient farm hauler is accessible in the market, numerous specialists and fabricates are not attempting to develop the innovation. Because of current advancement and future requests of cutting edge farm trucks, it is founded that around 800,000 work vehicles furnished with offices like farm truck direction will be traded in 2028, while a similar report supposes that around 41,000 automated, completely self-governing (level 5) farm trucks will be traded by starting of the 2040s. When discussing such refined machinery, most ranchers cannot bear to claim them, though a large portion of farm hauler specialist organizations and producers work well underneath their latent capacity. Thinking about the test, Hello Tractor has fostered an answer for sorting these issues. The organization has fostered a minimal expense observing gadget that can be put on any work vehicle and gives incredible programming and investigation instruments. The advantages of this gadget are twofold—on one side it guarantees that the general expense of the farm truck stays moderate for the vast majority of producers, while simultaneously it screens the state of the work vehicle and reports if any issues happen. The product associates work vehicle proprietors to ranchers needing farm hauler administrations, very much like Uber for farm vehicles. Another significant model is Case IH's Magnum arrangement farm hauler which utilizes on-panel camcorders and LiDAR sensors for object location and impact evasion. As of late, Case IH utilized this farm vehicle to sow soybeans by the subsequent idea of self-governing farm vehicles. In added advancement completed by norms bunch ETSI utilizing IoT to command the mishaps because of ranch vehicles. In the wake of gathering all the significant yield information, the next step is strident figuring from the cloud-based to the rim, such as John Deere needs.

In their proposed framework, an examination motor works locally on the rancher's farm hauler instead of in the cloud to change the nearby sources of info. For this reason, they considered all the current examinations and suggestions to alter the current information progressively relying upon the field conditions. In light of this marvel, the maker is carrying their work vehicles to the succeeding level by associating their appliance to the Internet and making out the strategy for showing the data any place rancher needs to see it. For example, here an IoT-based tractor is shown in Figure 3.3 which is effectively using sensors with a controlling board.

Figure 3.3 IoT-based tractor.

3.4.1.1 Event detection algorithm

To raise the period of wireless sensor network (WSN), the exponentially weighted moving average (EWMA) algorithm is accepted. The data dispatch to the cloud recurrently whensoever its rate drops down underneath or overhead the two predetermined thresholds. The main drawback of WSN is the power drawdown because of the transmission between sensor network nodes. Event Detection Algorithm is accepted to overawe the mentioned issue like cricket event prediction, etc. [8]. Many forms of the analysis covered in this existing literature are based on energy meter so it is very helpful to use in agriculture data or agriculture study and analysis because methods of the study and analysis are very technically matched so that work also as the future direction for researcher in agriculture field [9]. An exponentially weighted moving average is one of the threshold-based exploratory data analysis (EDA) which is implemented in the analysis, in the direction to establish the regulator limits of topsoil moisture. EWMA graph approach is a convenient method for identifying minor shifts of position. The measured representation of the algorithm is expressed as underneath:

$$EWMAi = Ei = \lambda ai + (1- \lambda)\ Ei - 1$$

aimed at $t = 1,2, 3\ldots, n$

where i = number of inspections at periodic intervals
λ = weighted mean supported the preceding values and lies between 0 and 1
a = obtained value

$Ei - 1$ = preceding value of Ei

$$\text{Upper Control Limit} = \mu 0 + Y\sigma$$

$$\sqrt{(\lambda/(2 - \lambda)[1 - (1 - \lambda)^{2i}])}$$

$$\text{Lower Control Limit} = \mu 0 - Y\sigma$$

$$\sqrt{(\lambda/(2 - \lambda)[1 - (1 - \lambda)^{2i}])}$$

where $\mu 0$ = Mean; Y = size of control limits; σ = standard deviation from the stirring range graph [10].

3.4.1.2 Design of the descending seed course monitoring (DSCM) appliance

The DSCM appliance is comprised of a tapered inlet and seed-control slot. This seed-control slot was planned for allowing descending seeds to travel through IR constantly. Furthermore, the tapered inlet could aid seeds to come in the seed-control slot evenly. While perceived in the diagram, a synchronize structure was put up on the center of the seeding plate. Whenever seed was released by seed plate, the intended course was found to be a parabola. So, in the study, the altitude between the entering of the seed-guiding slot and seed-releasing point was 42 mm. The course in X- and Y-directions can be calculated by the below equations:

$$X = w.r.t$$

and

$$h = 1/(2).g.t2$$

where X represents seed's dislocation in X-direction,

w is the angular velocity of seed plate
r represents the radius of the seed plate
t represents the time when the seed was dropped
h is descending seed's dislocation in Y-direction
g indicates the free-fall acceleration (9.8 m·s −1).

For avoiding any blocking process, the length of control slot L must be greater than the biggest width of the seed. Though, if the distance between the transmitter and collector was more, then the recipient may not get the IR ray continuously. So, the range of L should be set according to the calculated length. As to the leaning point ∅, in case it was set to more noteworthy than 60°, the inlet would turn out to be little and couldn't cover the entire outlet of the seed-metering gadget. However, if the leaning was set to a more modest 30°, the sensor would be cumbersome and hard to be joined onto the seed-metering gadget. Thus, the leaning point ∅ of the tightened divider was between 30° and 60°.

The width of the seed-control space additionally needed to be thought of. The width of the IR beam should be more modest than the seed's base measurement. Else, all things considered, more than one seed screened the IR beam at a case. Therefore, the breadth of the IR beam was set to not as much as the seed's base distance across. Alongside, the width of the control space D should fulfill the succeeding condition:

$$Lmax<D<Lmin$$

where Lmax and Lmin denote the most extreme and the base of the seeds' widths (in mm). Also, the D should be more prominent than the limit of the seed width Lmax. Or else, the seeds whose width is greater than D will be trapped at the entry of the seed control slot. This DSCM appliance has been shown here more clearly for seed falling in Figure 3.4.

3.4.2 Collecting robots

Reaping is the most basic stage during the creation cycle, as this last stage directs the yield and, eventually, its prosperity. In certain harvests, this is

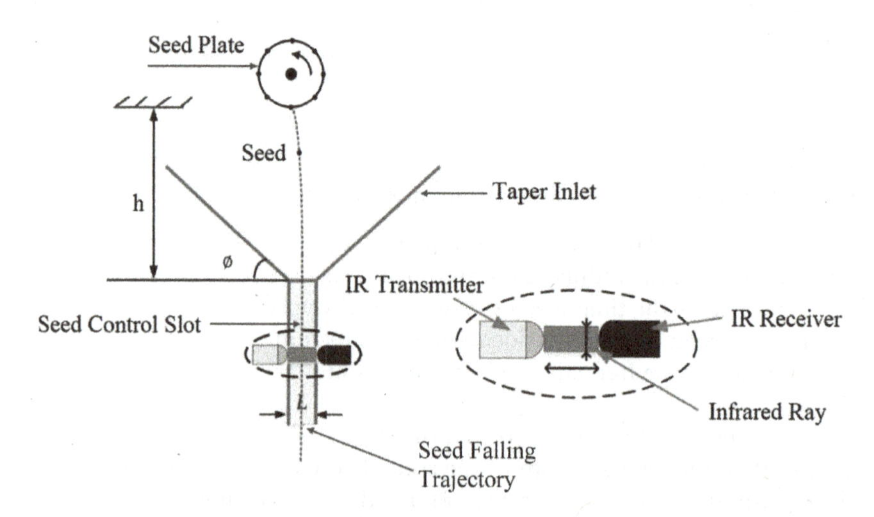

Figure 3.4 The descending course whenever the seed drops through the seed-control slot.

done a solitary time while, in some others, played out a few times, even consistently, as yield arrives at a specific stage. Reaping the yield at an opportune while is extremely basic, as deed so either initial or late can influence the creation essentially. When discussing the work, it is assessed that the United States airs a $3.2 billion decrease in yield creation consistently because of work lack. However, as per an examination directed by the US Department of Agriculture, by and large, 15% of homestead prices go to pays and work prices, whereas it tends to be up to 40% in nearly working serious ranches. Seeing the value of this phase and work matters, ranch specialists assume that the contribution of farming advanced mechanics might facilitate the work pressure as well as give the adaptability to reap at whatever point required. To mechanize the gathering cycle and create it further exact, part of robots has been expanding over new many years. Since the robot administrations, numerous scientists have done concentrated exploration to develop the affectability of organic product discovery, its shape, size, shading, and limitation. Programmed gathering of organic products requires profound examination of refined sensors that are fit for gathering exact and unambiguous data of that specific harvest and natural product. The errand of distinguishing the correct objective in normal scenes isn't basic since the vast majority of the organic products are impeded in part—now and then even completely—under the leaves and branches or are covered with different natural products. Here, the vast majority of the conspicuous investigations found concerning this design are profoundly founded on PC vision, picture handling, and AI methods. This cycle needs exceptionally specific and complex instruments to separate the organic product circumstances, as there are more than sizes, shapes, and shadings for a pepper unaided when it is prepared to gather. Thinking about alike intricacy, numerous robots are being created for explicit yields. A segment of primary robots used for produce reaping incorporates SW 6010 Octonion and SW 6010 for fruits, SWEEPER robot for peppers, and Robot for tree-based natural items which can get to tens of thousands of natural products every hour.

3.4.3 Correspondence in agriculture

Correspondence and detailing the data on an ideal premise are viewed as the foundation of accuracy farming. The genuine reason can't be accomplished except if a firm, solid, and secure association among different taking an interesting object is given. To accomplish correspondence dependability, telecom administrators can assume a vital part in the rural area. On the off chance that we need to carry out IoT for an enormous scope in the agribusiness business, we need to give a reasonably huge design. Here, the components like expense, inclusion, energy utilization, and unwavering quality are basic and must be considered before picking the average of correspondence. Low-power organizations can give networks just on one spot and for the most part, don't offer administrations in distant regions where detected information should be communicated to the ranch the board framework (FMS). Contingent upon accessibility, adaptability and application prerequisites, different correspondence styles, and advancements are being utilized for the reason, and utmost normal are examined here.

3.4.3.1 Cellular communication

Cell correspondence means from 2G to 4G may be appropriate, contingent upon reason and transfer speed necessity; notwithstanding, the unwavering quality, and accessibility, of the phone system in country regions is significant apprehension. For handling this, information broadcast using satellite is a choice, yet, now, the expense of this correspondence method is exceptionally high, which makes it not appropriate for little and midmost ranches. The decision of correspondence method likewise relies upon application prerequisites, for example, a few farms needed sensors that can work with low-slung information frequency however required to work for significant gives subsequently appeal extended battery lifespan. For alike circumstances, the alternative scope of low-power wide area network (LPWAN) is observed as a better answer for cell accessibility, as far as extended battery lifespan and a greater network region with moderate charges ($2 to $15 each year). At present, harvest and farm are the two managers of the principal applications, whereas LPWAN networks are exceptionally suitable, and, farther thoughtful about its success, it very well might be utilized in many other farming-related employments.

3.4.3.2 Zigbee

Zigbee is principally intended for an extensive scope of uses particularly to supplant current nonstandard advances. The IoT based gadgets are contingent upon the application necessities, that can be customized using Controller, Router and User [11]. In light of these attributes, and further seeing the horticulture application prerequisites, Zigbee may assume an essential part particularly focusing on the nursery climate where normally short reach correspondences are required. During checking the different boundaries, the ongoing information from the detector hub is moved over Zigbee to the last worker. For the applications such as water systems and treatment, Zigbee elements are arranged in such a way to correspondence, for example, in dribble water system utilized to screen topsoil substances such as dampness. Further, SMS is sent to a rancher to refresh about field information where GSM is needed at a significant distance or Bluetooth element can aid at more limited distances.

3.4.3.3 Bluetooth

Bluetooth, a remote correspondence level that interfaces little main gadgets composed over more limited distances as a rule collaborating in a closeness. Due to its benefits of low-power necessities, ease of use, and least expensive, the invention is being used at numerous shrewd agricultural appliances. Further, its production improvements in abundant IoT backgrounds with the entrance of Bluetooth smart or Bluetooth low power (BLP). The inspection directed in which trials Bluetooth and PLC (programmable rationale regulator) with ICP (incorporated control procedure), clock controller, and topsoil moisture regulator method for the brilliant irrigation system. The focus of such examination is to track down a supreme usage of water and energy application for diverse nursery or farm applications. A moisture and temperature sensor reliant on BLP is shaped in mainly zeroing in on the agricultural conditions and climate conditions of yield farms. Here, the clarification of choosing BLE for correspondence reason for current is its trademark support for cutting edge cell receptiveness. Further, comparable exertion is implemented where another detector hub is intended to screen encompassing light and temperature utilizing the BLP correspondence convention ideal for IoT-based agribusiness applications. Along short reach, Wi-Fi is utilized at all points. LAN communications are needed at shrewd farming. The study introduced explores a distant checking framework using Wi-Fi, where the detector hubs rely on WSN802G components. The conveyed hubs confer remotely with a local worker, which is reliable to collect and stock the noticed info and further authorizing viewing of the data after mandatory inspection.

3.4.3.4 LoRa

Lora remote innovation is a long-range, low-energy stage utilized widely in IoT devices. Being low in energy utilization, it bids LPWAN network among the remote sensors and cloud. It has substantiated itself considerably further successful and solid in comparison to Bluetooth, Wi-Fi, and so on, particularly in cafe conditions. Detectors dependent on LoRa might be introduced in more modest gadgets meant for solid checking. In particular, LoRa signs can enter thick and protected items, even structures, and can, henceforth, cover a bigger organization region. Generally, LoRa-based organizations perform higher as far as life expectancy and, simultaneously, present diminished support and upkeep trouble. Thinking about the benefits, numerous specialists tried this specialized technique in kitchens, extra spaces, and transport frameworks. A trial was directed in the stockroom by an ability to stock 40 tons of apples, and outcomes demonstrated that it gave complete inclusion, where wind stream and temperature readings moved effectively with the parcel pace of nearly 95%. Additionally, it grants a framework to accomplish data recognizability in the cereal transportation framework to guarantee the food quality by checking the stickiness and temperature levels.

3.4.3.5 *Sigfox*

Sigfox is likewise implemented to give system availability administrations to low-fueled articles as IoT is needed. It depends on ultra-narrowband or narrowband innovation; subsequently, it varies the period of a transporter radio wave to encrypt information. Because of these attributes, it offers undeniable level execution, regardless of whether 100 sensors need to communicate information simultaneously, as examinations were done.

3.5 CLOUD COMPUTING

Exactness farming is presenting its latent capacity and advantages by enlightening agrarian activities concluded better information determined dynamically. Notwithstanding, to proceed with this achievement, exactness farming not just requires better innovation and devices to handle information proficiently yet in addition at a sensible expense such as being utilized to settle on-field choices effectively. For this reason, ranchers can utilize cloud administrations to get data from prescient investigation organizations so they can pick the right item accessible as indicated by their particular necessities. Distributed computing bids a superiority to ranchers for utilizing information-based stores which contain a fortune of data and encounters identified with cultivating rehearses just as on gear choices accessible in the market with the vital subtleties. As a rule, this shows up with master counsel from a varied scope of sources. Cell phone-based detectors that being utilized in different horticulture devices. (Alike instance, on cultivating and handling of rural items.) To make it further viable, the situation could be stretched out further to incorporate admittance to shopper data sets, supply chains, and charging frameworks. Most likely, moving towards cloud-based administrations propose chances to investigate progressions, yet it accompanies new difficulties, too. Initially, a tremendous scope of sensors is being created and utilized in exactness farming, every one of which has its information configuration and semantics. Furthermore, a large portion of the choice emotionally supportive networks are application-explicit while, then again, a rancher could be in requirement of getting to several frameworks for the particular device, for example, topsoil checking. On the whole both of these cases, the cloud-based decision support framework is required to deal with a variety of information and their arrangements yet additionally should have the option to design these configurations for various applications.

3.6 CONCLUSION

The concentration on shrewder, improved, furthermore, more effectual harvest developing strategies is fundamental in mandate to meet rising food call of the growing populace of the world notwithstanding the consistently decreasing arable field. The progress of innovative approaches of further developing harvest yield and overseeing, one can eagerly see as of now: innovation forestalled, pioneering fresher individuals accepting agriculture as a career, farming as a method for freedom from non-renewable energy sources, stalking the crop growth, protection and sustenance classification, corporations against cultivators, contractors, vendors, and consumers. This paper thoroughly examined this load of qualities and emphasized piece of various innovations, particularly IoT, to make farming shrewder and much competent to encounter forthcoming opportunities.

REFERENCES

[1] Z. Ünal. 2020. "Smart Farming Becomes Even Smarter With Deep Learning—A Bibliographical Analysis", IEEE Access, 8: 105587–105609.
[2] N. Bhargava, A. K. Sharma, A. Kumar and P. S. Rathoe. 2017. "An Adaptive Method for Edge Preserving Denoising", 2nd International Conference on Communication and Electronics Systems (ICCES), 600–604.
[3] N. Khan, R. L. Ray, G. R. Sargani, M. Ihtisham, M. Khayyam and S. Ismail. 2021. "Current Progress and Future Prospects of Agriculture Technology: Gateway to Sustainable Agriculture", Sustainability, 13(9). https://doi.org/10.3390/su13094883
[4] A. K. Sharma, A. Dhaka, A. Nandal, K. Swastik and S. Kumari. 2021. "Big Data Analysis: Basic Review on Techniques", in Advancing the Power of Learning Analytics and Big Data in Education, A. Azevedo, J. Azevedo, J. Onohuome Uhomoibhi and E. Ossiannilsson, Eds. Hershey, PA: IGI Global, 208–233.
[5] A. K. Sharma, A. Nandal, A. Dhaka and Rahul Dixit. 2020. "A Survey on Machine Learning-based Brain Retrieval Algorithms in Medical Image Analysis", Health and Technology, 10: 1359–1373.
[6] K. Gautam, V. K. Jain and S. S. Verma. 2020. "Identifying the Suspect Nodes in Vehicular Communication (VANET) Using Machine Learning Approach", Test Engineering & Management, 83(9): 23554–23561.
[7] A. K. Sharma, A. Nandal, A. Dhaka and R. Dixit. 2021. "Medical Image Classification Techniques and Analysis Using Deep Learning Networks: A Review", in Health Informatics: A Computational Perspective in Healthcare, R. Patgiri, A. Biswas and P. Roy, Eds. Singapore: Springer, 233–258.
[8] K. Kanhaiya, R. Gupta and A. K. Sharma. 2019. "Cracked Cricket Pitch Analysis (CCPA) Using Image Processing and Machine Learning", Global Journal on Application of Data Science and Internet of Things, 3(1): 11–23.
[9] A. K. Dubey, A. Kumar, V. G. Díaz, A. K. Sharma and K. Kanhaiya. 2021. "Study and Analysis of SARIMA and LSTM in Forecasting Time Series Data", Sustainable Energy Technologies and Assessments, 47: 1–14.

[10] M. C. Morais, S. Knoth and C. H. Weiß. 2019. "An ARL-unbiased Thinning-based EWMA Chart to Monitor Counts", Sequential Analysis, 37(4): 487–510.
[11] A. K. Sharma, K. Kanhaiya and J. Talwar. 2020. "Effectiveness of Swarm Intelligence for Handling Fault-Tolerant Routing Problem in IoT", in Swarm Intelligence Optimization: Algorithms and Applications, A. Kumar, P. S. Rathore, V. G. Diaz and R. Agrawal, Eds. Wiley Online Library: Wiley, 325–341.

Chapter 4

IoT and FOG space-time Particulate Matter ($PM_{2.5}$) concentration forecasting for IoT-based air pollution monitoring systems

Abirami Sasinthiran and Chitra Pandian

CONTENTS

4.1 INTRODUCTION

With the acceleration of urbanization, air pollution has become a major concern worldwide. It causes a severe risk to the environment's health. As per the reports released by the World Health Organization in the year 2016, air pollution has caused around 4.2 million premature deaths across the globe. Particulate Matter, ground-level ozone, carbon monoxide, sulphur oxides, nitrogen oxides, and lead are some of the air pollutants that have been contaminating the atmosphere from many sources. Among them, Particulate Matter ($PM_{2.5}$) is one that menace more on the health of human beings. $PM_{2.5}$ are extremely small air pollutant particles with a

DOI: 10.1201/9781003230236-4

diameter of less than 2.5µm. Prolonged exposure to $PM_{2.5}$ causes various acute and chronic effects on the respiratory and cardiovascular systems in humans [1]. People suffering from these disorders are highly susceptible to bad air quality with increased $PM_{2.5}$ concentrations and are called vulnerable groups.

Air quality monitoring is essential to monitor and safeguard a healthier environment for the future. The high cost of installing monitoring stations and the need for precision in observations make fine-grained air quality monitoring difficult to achieve [2]. IoT-based solutions to evade the ill effects of air pollution would be a more proficient approach to overcome this challenge [3]. One of them is IoT-based air pollution monitoring [4]. A significant number of IoT air quality sensors are put at various sites around an area in IoT-based air quality monitoring. These sensors collect a lot of data regarding air quality, including concentrations of various pollutants in the area. These air quality data, when accurately modelled, would yield useful knowledge that could likely benefit society and the environment. The patterns extracted from them can help in forecasting future air quality values. Forecasting ambient air values is extremely useful for vulnerable people when it comes to preparing for outdoor exposure [5]. Also, forecasting can warn the public sectors about future adversities in air quality, enabling them for precautionary measures.

The nature of air quality data is immensely non-linear, chaotic, and randomly changing [6]. Deep learning techniques that use several non-linear transformations to air quality data can provide higher level abstractions [7]. Data on air quality acquired from various sensors at various places is spatiotemporal, implying it has both spatial and time properties. The ability to define and relate spatiotemporal characteristics in data is a significant topic of research in artificial intelligence with numerous applications in the field of Geographic Information Systems [8]. Inter-dependency (spatial autocorrelation) and intra-dependency (heterogeneity) features are present in all spatiotemporal data [9]. The inter-dependency attribute asserts that a location's value varies based on its surrounding values. For instance, if the air quality in the vicinity of a site 'A' is poor (due to trash burning, building excavation, etc.), the air quality in the location 'A' will be poor as well. The intra-dependency attribute asserts that the spatiotemporal data is often not evenly distributed. The value of air quality is determined by the time of day when the observed instant occurs. For instance, due to continuous traffic, the air quality of a place 'A' is anticipated to be poorer in the morning than in the afternoon.

Many researchers aim to model and forecast the air quality values considering it as a plain regression problem. Typically, classical data mining approaches [10–12] and machine learning approaches [13, 14] were used for forecasting without considering the spatiotemporal relations in the data. This led to unsatisfactory performance. Recently many deep learning algorithms for forecasting consider the spatiotemporal nature of the data. But they all trail behind in overcoming two challenges: (1) well synchronizing the association between the identified spatiotemporal relations with the forecast and (2) tracing much longer spatiotemporal dependencies that prevail in the data. In this paper, the proposed architecture takes advantage of a hybrid CNN-LSTM architecture, which is an extension of our earlier proposed architecture [15], to overcome the above-mentioned two challenges. The air quality data is initially transformed into an image-like data to preserve the spatial information in it. Hence, the air quality forecasting problem is formulated as an image prediction problem. This provides ease and accuracy in identifying the inter-dependencies in the data. The 1D-CNN employed for capturing the heterogeneity in every 24 hours of data shows beneficence for long short-term memory (LSTM) to capture much longer dependencies in the data.

The following are the paper's contributions: (1) Convert air quality data to an image-like format to emphasize the data's true spatial distribution; (2) Employ 3D-CNN and 1D-CNN to capture inter- and intra-dependency in every 24-hour air quality data; (3) Implement LSTM to perform sequence learning on encoded data and forecast future spatiotemporal relationships; and (4) Demonstrate the suggested methodology's efficacy by means of prediction accuracy and computing cost.

4.2 IOT- BASED AIR QUALITY MONITORING

IoT architectures are intended to provide solutions in a wide range of applications [16]. The fundamental three-layered IoT model for air pollution monitoring comprises the Application layer, the Network layer, and the Sensor layer as revealed in Figure 4.1.

The air pollutant sensors present at the Sensor layer collect the concentration of various air pollutant and send them to the Network layer. The Network layer acts as a middleware between the Sensor layer and the Application layer. The air quality data from the Network layer, which includes various air pollutant concentrations, is broadcast to the Application layer. The air quality data received at the Application layer is investigated and analysed by the services provided by the Application layer. This results in the extraction of useful knowledge and patterns that underlie the air quality data. This useful knowledge extracted from the air quality data benefits the vulnerable group and air pollution control management who are the end users of the Application layer. Our proposed work is intended for the extraction of useful patterns and predicts future air qualities in the Application layer of IoT-based air quality monitoring architecture.

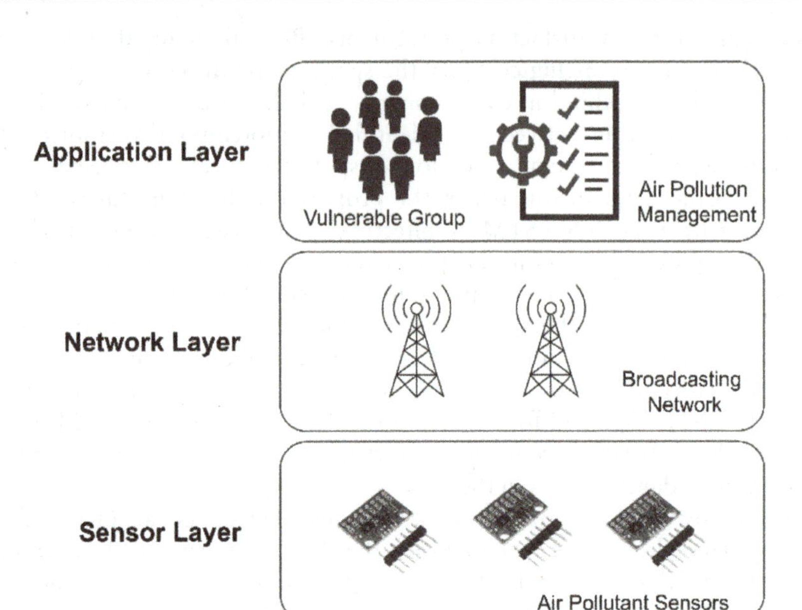

Figure 4.1 Three-layered IoT model for air pollution monitoring.

4.3 RELATED WORKS

Starting from physical and mathematical models, research on predicting air quality values and air pollutant concentrations has a long history [17, 18]. These methods use complicated prior information to model the spread of air contaminants. Later artificial intelligence knowledge-based techniques sprouted out for environmental decision support systems [19]. Some of the shallow learning models proposed for air quality prediction include multiple linear regression by Li et al. [20], the autoregressive moving average (ARMA) by Box and Jenkins [21], and the support vector regression (SVR) method by Nieto et al. [22]. For extracting spatiotemporal correlations in air pollution data, Qin et al. [23] presented an apriori-pattern mining technique. Probabilistic graphical models are also explored for air quality [24] and air pollutant forecasting [25]. In addition, numerous variations of kriging, a geostatistics approach for prediction, have been employed to improve air quality forecasts. Among these are Markov Cube-Kriging by Liang and Kumar [10], kriging with external-drift by Pearce et al. [11], and Bayesian-kriged Kalman-filtering by Sahu and Mardia [12]. They've all successfully predicted the nonstationary spatiotemporal structure of air pollution data. A combined framework joining neural and neuro-fuzzy modules for air quality prediction by Neagu et al. [26] incorporated e knowledge acquired from human experts and produced encouraging results. For monitoring the $PM_{2.5}$ value, Gupta and Christopher [14] built a neural network capable of simulating the non-linear and interaction relationship in air quality data. Ghaemi et al. [13] designed a online spatiotemporal system named LaSVM for huge and streaming data using a modified support vector machine. For real-time air quality monitoring and evaluation, decision trees, artificial neural networks, and support vector machines have also been investigated [27]. Multitask learning, which investigates commonality among tasks to tackle several learning tasks for enhanced predictive performance for regular concentration forecasting of airborne pollutants, has been investigated by Zhu et al. [28].

In today's world, deep learning algorithms find a wide range of applications in various fields. Deep learning techniques are being explored as a follow-up to machine learning algorithms for modelling the complicated non-linear interactions of spatiotemporal air quality data [29]. Shengdong Du et al. [30] developed an integrated methodology for $PM_{2.5}$ prediction using 1D-CNN and BiLSTM. For wind speed prediction, Qiaomu Zhu et al. [31] built a unified architecture combining a CNN and a LSTM. This model was trained end-to-end with a unique loss function, which learns both temporal and spatial correlations. For modelling air quality data, Duc Le et al. [32] suggested a CNN-LSTM architecture to analyse image-like spatial and temporal data. Huang and Kuo [33] created APNet for $PM_{2.5}$ forecasting, which took into account both meteorological and pollutant data. Seonggu Lee and Shin [34] used a combination of convolutional LSTM and CNN to predict $PM_{2.5}$ concentrations with greater accuracy. To increase predicted performance, the technique took into account a variety of auxiliary data, including spatial information. Air pollutant $PM_{2.5}$ forecasting by Qin et al. [35] and ozone concentration forecasting by Pak et al. [36] have also exploited different hybrid combinations of CNN and LSTM for deep learning of the features in the data. Apart from air quality, hybrid combinations of CNN with variants of LSTM have also been proposed for various other applications such as typhoon formation forecasting [37], Spam detection in social media [38], and malicious web page detection [39].

All these CNN-LSTM architectures mentioned above focus on forecasting by capturing the spatiotemporal features in the data. They lack to consider the much longer dependencies in the data that are lost during the sequential gradient propagation through deep layers of LSTM. Considering these much longer dependencies adds an advantage to improved prediction accuracy. With this motivation, we propose a hybrid CNN-LSTM architecture to extract substantial spatiotemporal features within the air quality data to forecast at reduced computational complexity.

4.4 OVERVIEW

The description of the data used for the proposed study and the problem formulation of the proposed work is as discussed below.

4.4.1 Data description

The proposed work's goal is to model the real-world air quality data set named IoT_City_Pulse_Pollution data set to forecast $PM_{2.5}$ concentrations.

This data set contains the concentration values of five air pollutants namely $PM_{2.5}$, carbon monoxide, nitrogen dioxide, ozone, and sulphur dioxide that were collected at an interval of five minutes. These air pollutant concentrations are the features considered for forecasting. The air pollutant concentrations were collected using IoT sensors placed at 449 locations. In this work, considering the training time we model the air pollutant concentration values collected at 25 locations that approximately fall on a 0.1° × 0.1° grid as shown in Figure 4.2. Each grid is called a cell and is denoted as c_i where $i \in \{0,24\}$. Table 4.1 shows the details of the data set under consideration.

The intra-dependency (heterogeneity) of a time series data can be visualized by calculating its autocorrelation function. The autocorrelation of $PM_{2.5}$ data at Cell 0 is as shown in Figure 4.3(a). Figure 4.3(a) shows significant autocorrelation which implies that the air quality data is a non-stationary time series with strong temporal correlations existing in it. Similarly, the spatiotemporal dependency between two cells can be illustrated through the Spatiotemporal Morans's I [40] calculated between $PM_{2.5}$ data of the cells.

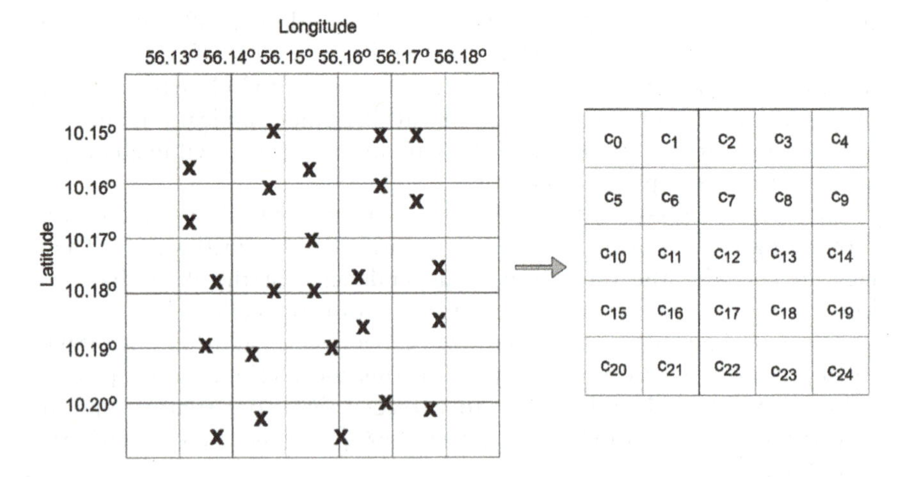

Figure 4.2 Representation of cell.

Table 4.1 Details of the considered data set

Data set name	*IoT_City_Pulse_Pollution*
Region	Aarhus
Number of considered stations	25
Duration	1/8/2014 to 1/10/2014
Number of variables	5
Interval considered	5 minutes
Total entries	17,568 entries for each location

The Spatiotemporal Morans's I value between all cells is given as revealed in Figure 4.3(b). The heat map in Figure 4.3(b) depicts that there exist strong spatial dependencies between a cell and its near cells (neighbours) and the least spatial dependencies between a cell and its far cells.

4.4.2 Problem formulation

Initially, the data at a particular time instant is transformed into an image-like structure. In this image-like data representation, we impose each location to represent a pixel, and every feature information for all nine locations at a time instant considered for forecasting represents a band of an image. The neighbourhood details of the locations are preserved through this transformation. After transformation, the entire data for all nine locations considered at all time instants would appear as 3D tensors distributed over time as shown in Figure 4.4. As we consider nine locations, we have a 3 × 3 image structure for every pollutant (feature) which represents a band of an image. As we have five features, the value of $n = 5$.

The considered entire data is characterized as a set of real numbers denoted as $D \in \mathbb{R}^{T x m x n}$, where 'T' denotes the total time instants, 'm' denotes the total locations considered, and 'n' denotes the number of features considered. After image-like data transformation, at every time instant say 't', we have a tensor of values represented as $G(f,s,t)$ where f represents a feature, $f \in$ {ozone, carbon monoxide, sulphur dioxide, nitrogen dioxide, $PM_{2.5}$}; s represents location, $s \in \{0,24\}$; and t represent every time instant separated by five minutes interval. The entire data is divided into groups of consecutive 24 hours data (288 time instants), each group represented as 'P'. For example,

$$P_1 = \{G(f,s,0), G(f,s,2), \ldots\ldots G(f,s,287)\};$$

$$P_2 = \{G(f,s,288),\ G(f,s,289), \ldots\ldots G(f,s,575)\} \text{ and so on.}$$

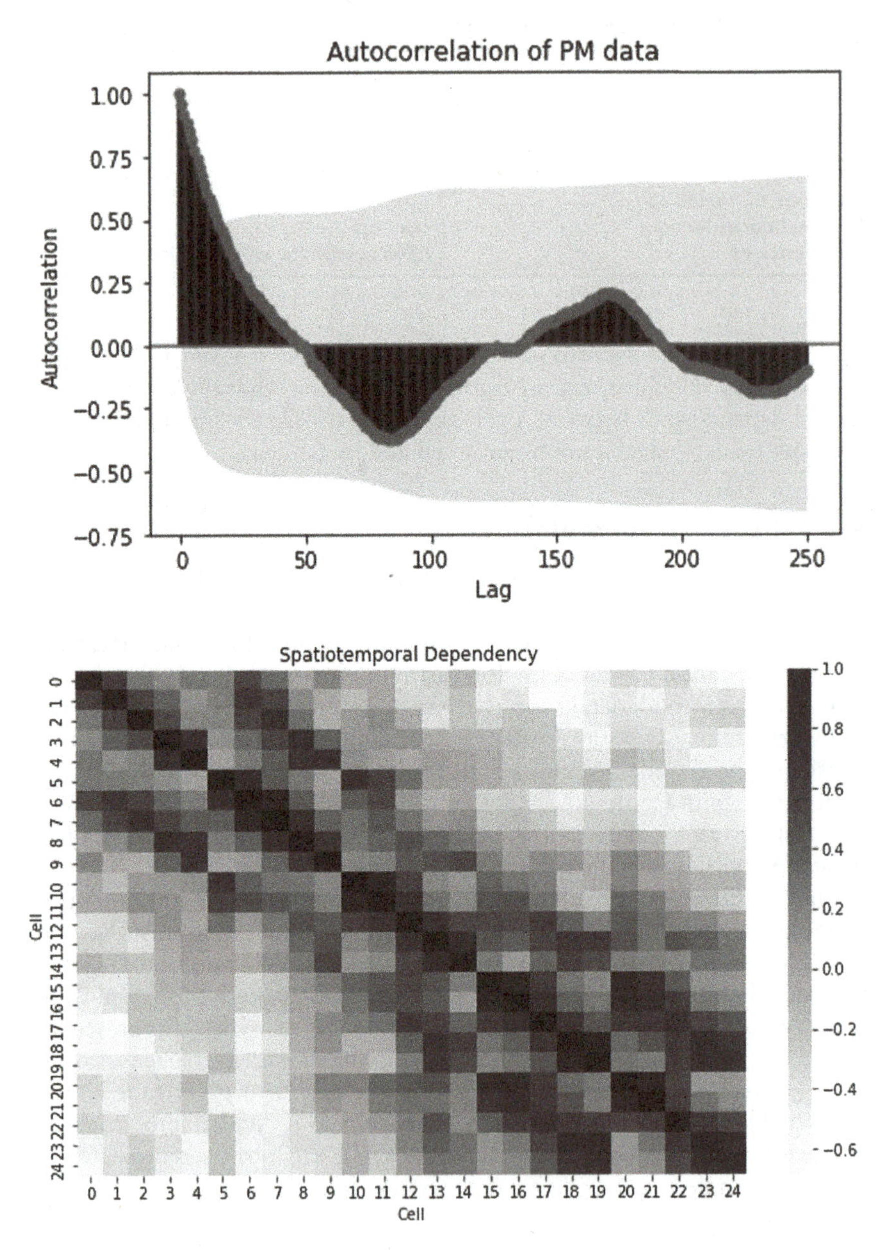

Figure 4.3 (a) Temporal correlation in $PM_{2.5}$ data of cell 0; (b) Spatiotemporal correlation between $PM_{2.5}$ data of all cells.

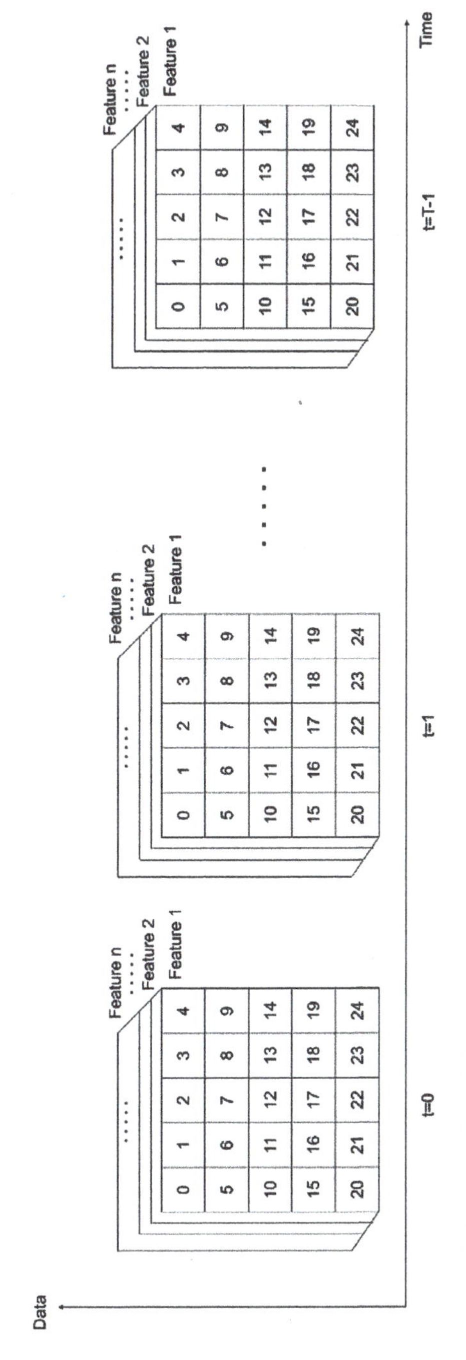

Figure 4.4 Time-distributed 3D-input structures.

Let 'Enc' refer to the operation of encoding spatiotemporal dependencies in 24 hours data. The aim of the proposed work is to pin down a function 'F' that satisfies the following:

$$\tilde{P}_{N+1} = F\left(Enc(P_1), Enc(P_2), Enc(P_3), \ldots Enc(P_N)\right) \tag{4.1}$$

where $\tilde{P}_{N+1}$ represents the forecasted values for the next 24 hours and $N = T/288$. The proposed architecture is intended to perform the function 'F' with higher prediction accuracy.

4.5 PROPOSED METHODOLOGY

The proposed methodology includes two main units namely the CNN-based encoder and LSTM unit, which are as discussed below.

4.5.1 Proposed CNN-based encoder

As mentioned in Section 4.4.2, the data is first preprocessed for outliers and then turned into image-like data. The proposed CNN-based encoder receives the modified data as input. The related inter-dependency and intra-dependency features in the transformed image-like input of 25 locations during a 24-hour period are extracted using CNN, which is commonly utilized for applications that process images. Therefore, the output of the encoder depicts the distribution of air pollutants over a region consisting of 25 locations varies concerning their neighbours within a day duration. Further, the LSTM processes the encoded spatiotemporal dependencies in the latent space for series learning. Figure 4.5 illustrates the proposed CNN-based encoder.

The 3D-CNN captures the spatial relations among the 25 locations at every time instant. The 1D-CNN added in the encoder captures the heterogeneity in 24 hours data. The computational cost for sequence learning using LSTM layers is $n \times d^2$, where 'n' denotes the sequence length and 'd' denotes the LSTM hidden state matrix. Without 1D-CNN, the sequence length 'n' to the LSTM layer is n = T. With 1D-CNN, the value of 'n' gets reduced to n = T/288. Hence, 1D-CNN adds benefit in reducing the computational cost. Also, with a reduction in sequence length, the amount of effort required by LSTM for series learning is reduced. As a result, the likelihood of LSTM failing to trace much longer dependencies in the data is minimized. This adds support for higher prediction accuracy.

4.5.2 Proposed LSTM for learning patterns in the series represented in latent space

The LSTM algorithm is notably good at learning sequential input and projecting future values. There are T/288 encoded data instants in the CNN output.

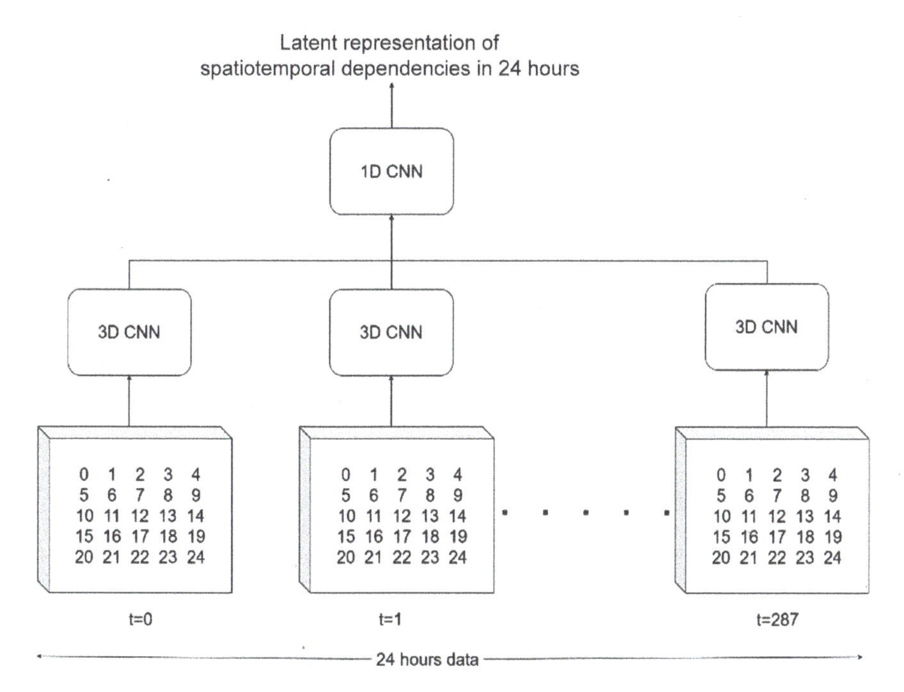

Figure 4.5 Proposed CNN-based encoder.

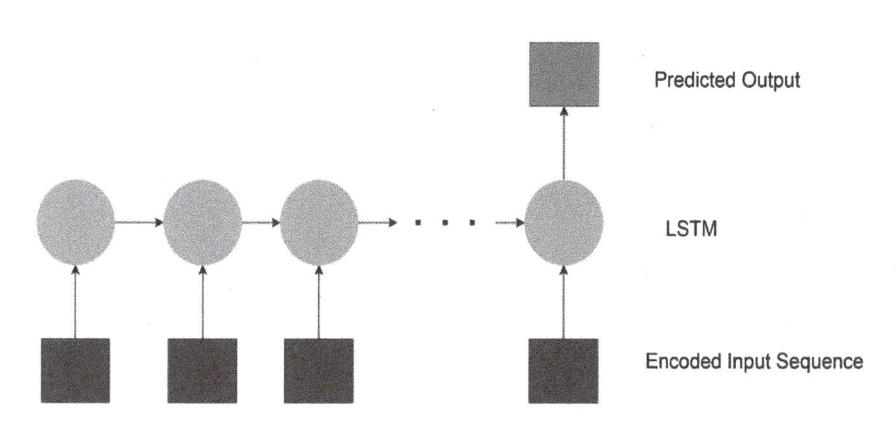

Figure 4.6 LSTM – many-to-one arrangement.

This T/288 encoded input sequence is fed into a set of LSTM cells placed in a many-to-one manner as illustrated in Figure 4.6 for anticipating the future 24 hours value for the 25 locations. Using the anticipated values in conjunction with the preceding inputs, as shown in Figure 4.7, many additional multistep future outputs can be obtained.

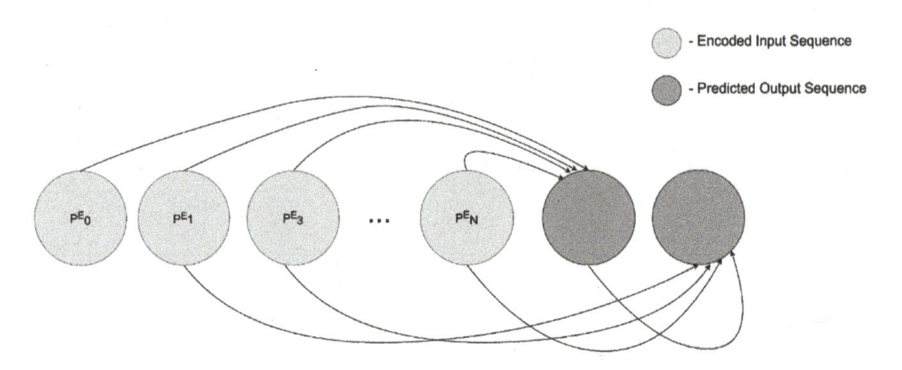

Figure 4.7 LSTM forecasting future $PM_{2.5}$ concentration.

4.5.3 Modelling

The image-like transformed data is fetched as input to the proposed CNN-based encoder in groups having 288 consecutive time instants (one-day) data per group. The proposed CNN-based encoder initially employs a 3D-CNN to retrieve the inter-dependency features in the 24 hours data. Furthermore, 1D-CNN captures a site's intra-dependency aspects, such as the day-to-day variation in air quality. The time distributed image-like data given as $\{G(f,s,0), G(f,s,1), \ldots, G(f,s,T-1)\}$ is split into groups each having consecutive 24 hours data (288 time instants). Each set of 24 hours group is represented as $\{P_1, P_2, \ldots, P_N\}$. The CNN-based encoder receives the time-distributed input in groups of 24. The encoder's output is represented as

$$Enc(P_i) = P_i^E = \left(C_1(C_2(P_i))\right), \text{where } i = 0,1,\ldots N \tag{4.2}$$

C_1 is the weight matrix evolved and progressed for 1D-CNN, whereas C_2 is the weight matrix evolved and progressed for 3D-CNN. The LSTM layers are used to learn the patterns in the sequence of encoded inputs, as seen in Figure 4.8.

Latent variable of an LSTM cell represented as 'Z' for a time instant is obtained through a set of non-linear operations (g) over the preceding instant's latent variable and the current instant's encoded input. It can be written as,

$$Z_{i+1} = g\left(Z_i, P_{i+1}^E\right), \text{where } i = 0,1,\ldots N \tag{4.3}$$

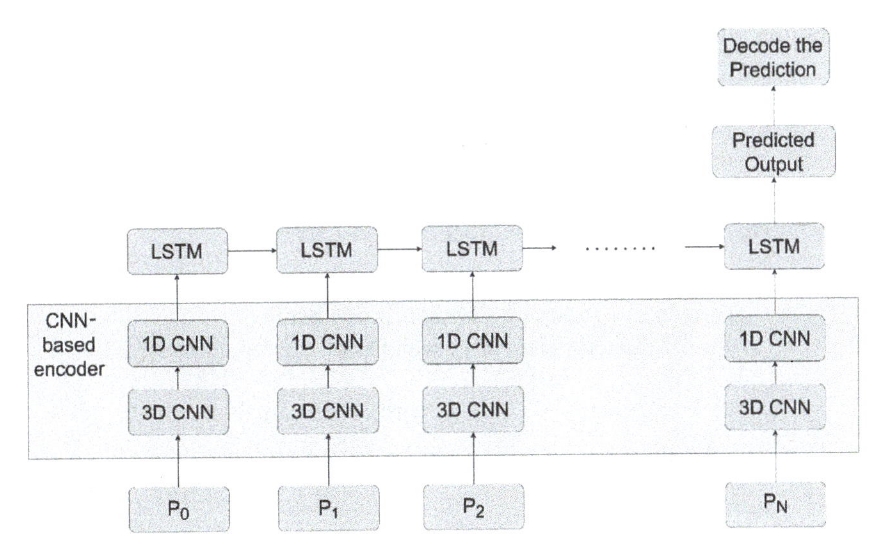

Figure 4.8 Proposed hybrid CNN-LSTM framework.

A non-linear operation (d) over the $N + 1$th latent variable (Z_{N+1}) yields the encoded forecast, that is, the encoded output of the $N + 1$th instant which is given as,

$$\tilde{P}^{E}_{N+1} = d(Z_{N+1}) \tag{4.4}$$

The proposed model's learning objective is to determine the non-linear functions g, d, C_1, and C_2 that minimize the proposed hybrid architecture's loss function, which is given as follows,

$$L(C_1, C_2, g, d) = \frac{1}{N}\sum_{t}\left(P^{E}_{N+1} - \tilde{P}^{E}_{N+1}\right)^2 \tag{4.5}$$

where P^{E}_{N+1} is the encoded value of the actual $N + 1$th batch in the set of 24 hours group. $\tilde{P}^{E}_{N+1}$ is the encoded forecast value which is the output of the $N + 1$th batch in the set of 24 hours group. The proposed model's learning objective is met using Stochastic Gradient Descent algorithm that computes C_1^*, C_2^*, d^* and g^* so that Equation (4.6) is satisfied.

$$C_1^*, C_2^*, g^*, d^* = \arg\min_{C_1, C_2, g, d} L(C_1, C_2, g, d) \tag{4.6}$$

As shown below, the encoded forecast output obtained is suitably decoded using the inverse operation of C_1 and C_2.

$$\hat{P}_{N+1} = (C_2^I (C_1^I (P_{i+1}^E)) \tag{4.7}$$

Algorithm 1:Training of proposed CNN-LSTM

Given: Input $\{P_1, P_2, \ldots, P_N\}$ a tuple of image-like transformed data grouped in batches of 24 instants per batch.
Initial values for the entire weights of the model to be learned $W = \{C_1, C_2, g, d\}$.
$\nabla \rightarrow$ Loss function's gradient with respect to W

Output: W (learned model)

1. while $\nabla \neq 0$ do
2. for every $i = 0$ to N do
3. Compute P_i^E using Equation (4.2)
4. end for
5. Compute the latent value of Z_{N+1} using Equation (4.3)
6. Compute the predicted encoded output $\hat{P}_{N+1}^E$ using Z_{N+1} as given in Equation (4.4)
7. Compute the loss as given in Equation (4.5) and its corresponding ∇
8. Update weights of C_1, C_2, g, d with respect to the computed ∇
9. end while
10. return W

where C_1^I and C_2^I represent the inverse operation of C_1 and C_2, respectively.
The proposed CNN-LSTM architecture is trained using Algorithm 1.

4.6 EVALUATION METRICS

The below-mentioned performance measures are used to evaluate the efficacy of the proposed hybrid architecture:

a. **RMSE – Root mean square error:** The RMSE is a metric that measures the difference between anticipated and real data. An efficient model is expected to have reduced RMSE value.

$$RMSE = \sqrt{\frac{1}{N}\sum_{i=1}^{N}(X_i - \tilde{X}_i)^2} \tag{4.8}$$

b. **MAE - Mean absolute error:** The MAE is a metric that measures the amplitude of errors in a collection of forecasts. Reduced MAE value indicates that the model's forecasts are reliable.

$$MAE = \frac{1}{N}\sum_{i=1}^{N}\left|X_i - \tilde{X}_i\right| \tag{4.9}$$

c. **R^2 - R squared:** R^2 is a metric that determines the degree of agreement between predicted and actual values. Therefore, higher RAE value implies that the predictions by the model are accurate.

$$R^2 = 1 - \frac{\Sigma(X_i - \widetilde{X}_i)^2}{\Sigma(X_i - \overline{X})^2} \tag{4.10}$$

where X_i is the actual value, $\widetilde{X}_i$ is the predicted value, and $\overline{X}$ is the average value of X.

4.7 EXPERIMENTS AND RESULTS

This section describes the experimental configurations and relevant parameters used for model validation. Keras, an open-source deep learning library based on Tensorflow, was used to develop the model. The proposed hybrid model is implemented using the 10th-Gen Intel Core-i5-7200U processor with 2GB-NVIDIA Geforce_MX250_Graphics_GDDR5. The whole data set is divided into 289 groups, each having 59 days of data (16,992 consecutive time instants) as input and immediate next day data (288 consecutive time instants) as output. Out of the 289 groups, 70% was used to train the proposed model and the remaining 20% was used to test it. Hence, the proposed CNN-LSTM architecture was trained with data in batches of 17,280 (60 × 288) instants for 200 epochs. The proposed hybrid framework is compared with the subsequent baseline models: (1) SVR, a kernel-based approach of machine learning, is used with radial basis function kernel and the epsilon value set to 0.1. (2) ARMA, which makes use of two polynomials namely moving average and autoregression to describe any stochastic and weakly stationary time series. (3) CNN, a deep learning technique for extracting hidden abstraction, consists of 3, kernel size-3, 1D-convolution layers with {256,128,64} filters, and pooling layers with ReLU as the activation function. (4) LSTM, a deep learning approach for handling sequences, involves {64,64} encode-decoder layers of LSTM cells and tan*h* as the activation function. (5) ConvLSTM, a variant of LSTM with internal matrix multiplications replaced by convolution operations, has a 2D kernel size set as [3,3] and tan*h* as the activation function. Shallow learning models such as

SVR and ARMA were implemented using Scikit-learn. The shallow learning models SVR and ARMA are made to predict a single location's $PM_{2.5}$ concentration value. The training parameters, batch size, and epochs size have values of 64 and 100, respectively. The proposed framework's performance in predicting $PM_{2.5}$ concentrations is evaluated by calculating the RMSE value, MAE value, and R^2 value of the model based on the predictions made during testing.

4.7.1 Performance comparison with baseline models

The comparison of RMSE value, MAE value, and R^2 value of the proposed approach with the baseline approaches are shown in Tables 4.2, 4.3, and 4.4.

Based on Tables 4.2, 4.3, and 4.4, we can infer that deep learning methods outperform shallow learning methods like SVR and ARMA. This highlights the importance of consideration of spatiotemporal relations during forecasting. In addition, we can find the proposed CNN-LSTM to have lower RMSE value and MAE value and significantly greater R^2 value while predicting $PM_{2.5}$ concentrations than the deep learning-based baseline approaches. The proposed CNN-LSTM outperforms the baseline

Table 4.2 Comparison of RMSE of proposed CNN-LSTM with baseline approaches

	Test					
Model	*#1*	*#2*	*#3*	*#4*	*#5*	*Average*
SVR	77.8	81.3	85.6	71.2	79.4	79.06
ARMA	67.8	75.4	86.9	81.3	74.4	77.16
CNN	57.6	61.2	55.6	49.5	56.2	56.02
LSTM	56.7	63.2	57.8	48.6	51.9	55.64
ConvLSTM	38.9	43.3	27.6	39.5	40.8	38.02
Proposed CNN-LSTM	24.2	31.3	18.9	27.6	25.3	25.46

Table 4.3 Comparison of MAE of proposed CNN-LSTM with baseline approaches

	Test					
Model	*#1*	*#2*	*#3*	*#4*	*#5*	*Average*
SVR	66.3	57.8	61.8	55.4	64.7	61.2
ARMA	51.2	48.9	53.2	45.5	56.7	51.1
CNN	38.9	44.5	39.2	48.9	47.4	43.78
LSTM	36.2	46.7	47.7	33.3	38.9	40.56
ConvLSTM	27.6	22.9	34.2	25.3	31.3	28.26
Proposed CNN-LSTM	26.1	19.7	17.4	22.7	24.2	22.02

Table 4.4 Comparison of R^2 of proposed CNN-LSTM with baseline approaches

	Test					
Model	*#1*	*#2*	*#3*	*#4*	*#5*	*Average*
SVR	0.69	0.73	0.67	0.71	0.72	0.70
ARMA	0.72	0.78	0.73	0.71	0.75	0.74
CNN	0.76	0.81	0.79	0.83	0.77	0.79
LSTM	0.81	0.77	0.84	0.76	0.85	0.81
ConvLSTM	0.88	0.77	0.85	0.79	0.89	0.84
Proposed CNN-LSTM	0.95	0.93	0.89	0.96	0.91	0.93

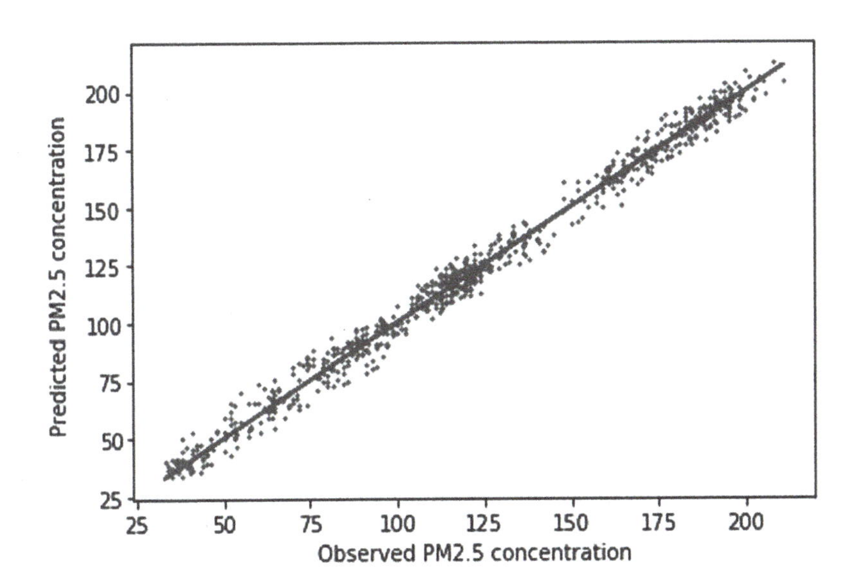

Figure 4.9 Predicted and observed data from the test set.

approaches with nearly 68% reduced RMSE value and 64% reduced MAE value. The proposed CNN-superior LSTM's performance is due to its effectiveness in explicitly capturing the inter-dependencies and intra-dependencies features in spatiotemporal relations and their influence on predicted results. As a result, the proposed CNN-LSTM forecast values are said to closely match the ground truth. Further, it is also verified by comparing the predicted outputs with the ground truth as illustrated in Figure 4.9.

The proposed CNN-LSTM predicts $PM_{2.5}$ concentrations that are consistent with the observed data, as shown in Figure 4.9. In addition, the proposed CNN-LSTM was also evaluated for assessing its performance in predicting multistep ahead predictions. Its performance comparison for

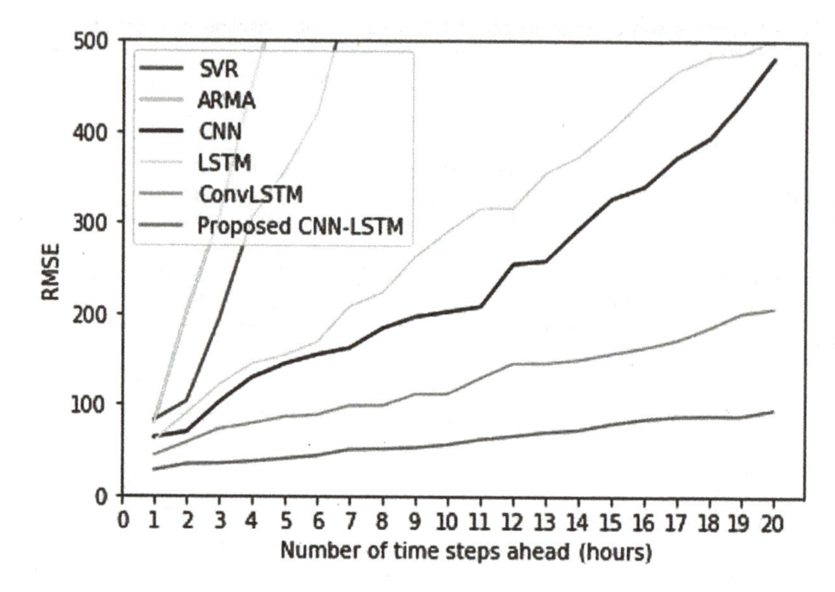

Figure 4.10 RMSE comparison for predicting multistep ahead instances.

multistep ahead predictions with the baseline approaches is illustrated in Figure 4.10.

When compared to the baseline techniques, Figure 4.10 indicates that the proposed CNN-LSTM has a lower RMSE value for predicting multistep ahead instances. The shallow learning approaches show a larger deviation in truth while predicting multistep ahead instants due to its inability to track the spatiotemporal trend in the data. Among the deep learning approaches, the proposed CNN-LSTM shows consistent and reduced deviation in truth while predicting multistep ahead instants. This is due to the ability of the proposed CNN-LSTM in preserving the dynamicity of the multi-time series during the encoding and latent representation process of prediction.

4.7.2 Effectiveness of 1D-CNN in the proposed CNN-LSTM architecture

The efficiency incorporated by 1D-CNN to the proposed architecture is illustrated by comparing the computational complexity and ability for multistep ahead prediction with and without its presence. Table 4.5 shows how the proposed architecture with and without 1D-CNN performs in terms of training and testing time.

Table 4.5 Comparison of computational complexity of proposed CNN-LSTM with and without 1D-CNN

	Training time (seconds)	*Testing time (seconds)*
Proposed CNN-LSTM without 1D-CNN	12,138.452	436.782
Proposed CNN-LSTM with 1D-CNN	4,008.695	80.934

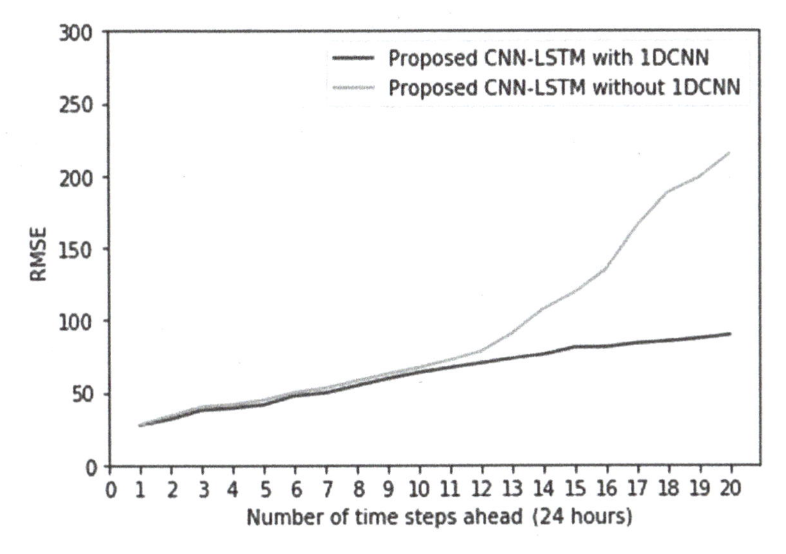

Figure 4.11 Comparison of proposed CNN-LSTM with and without 1D-CNN for multistep ahead prediction.

The number of LSTM in the hidden layer increases by 288 times without 1D-CNN. The computational complexity involved in learning through a larger number of multiplicative gates and recurrent connections of LSTM is higher than the convolution operations in 1D-CNN. This is the reason for the increased training time and testing time of the architecture without 1D-CNN in Table 4.5. The proficiency of both architectures for predicting multistep ahead instances is exhibited in Figure 4.11. Initially, both the architectures show the same RMSE values but after $y = 13$ the proposed CNN-LSTM with 1D-CNN shows better performance than the one without it. This signifies the role of 1D-CNN in supporting LSTM for capturing much longer dependencies in the air quality data.

From Sections 4.7.1 and 4.7.2, we infer that the proposed CNN-LSTM outperforms all baseline techniques in $PM_{2.5}$ concentration prediction, with the potential to capture considerably longer trends in air quality data. The predictions made by the proposed approach match well with the observed data while predicting both single step and multistep ahead instances. This demonstrates the proposed CNN-LSTM framework's potential to effectively learn local spatial trends as well as long-term temporal trends in air quality data set which is a multivariate time series. Hence, the proposed approach can be effectively utilized to provide accurate forecasts for early warnings to vulnerable groups and air pollution control management.

4.8 CONCLUSION

The need for air quality monitoring remains vital with massive urbanization occurring globally. IoT-based architectures provide an optimal solution for air quality monitoring. Along with monitoring, it also delivers real-time air quality analysis and forecasting for any location. The air quality values collected through various IoT sensors are stochastic and spatiotemporal. Deep learning algorithms proficiently extract the spatiotemporal patterns underlying the data set on air quality. Hence, deep learning techniques are rigorously employed for forecasting in IoT-based air quality monitoring architectures. In this paper, we propose a hybrid CNN-LSTM architecture for capturing all inter- and intra-dependency features in the data set on air quality for forecasting $PM_{2.5}$ concentration. The encoder–decoder architecture initially encodes all inter-dependencies and intra-dependencies in the data and later performs sequence learning over the encoded values for future prediction. The 1D-CNN in encoder improves the efficacy of the proposed architecture to capture much longer dependencies in the data paving the way for improved prediction performance. The proposed architecture was extensively evaluated with the IoT_City_Pulse_Pollution data set. The proposed CNN-LSTM effectively outperforms the baseline approaches with nearly 68% lower RMSE and 64% lower MAE values due to its ability to capture all spatiotemporal dependencies in the data. In the future, approaches based on generative deep learning shall be investigated for improved performance.

REFERENCES

[1] K.-H. Kim, E. Kabir, and S. Kabir, "A review on the human health impact of airborne particulate matter," *Environ. Int.*, vol. 74, pp. 136–143, 2015.

[2] C. Xiaojun, L. Xianpeng, and X. Peng, "IOT-based air pollution monitoring and forecasting system," in *International Conference on Computer and Computational Sciences (ICCCS)*, 2015, pp. 257–260.

[3] P. Chitra and S. Abirami, "Leveraging fog computing and deep learning for building a secure individual health-based decision support system to evade air pollution," in *Security, Privacy, and Forensics Issues in Big Data*, IGI Global, 2019, pp. 380–406.

[4] A. Khiat, A. Bahnasse, J. Bakkoury, M. El Khaili, and F. E. Louhab, "New approach based internet of things for a clean atmosphere," *Int. J. Inf. Technol.*, vol. 11, no. 1, pp. 89–95, 2019.

[5] P. Chitra and S. Abirami, "Smart pollution alert system using machine learning," *Integr. Internet Things Into Softw. Eng. Pract.*, vol. 56, pp. 219–235, 2018, doi:10.4018/978-1-5225-7790-4.ch011

[6] H. Tong, "A personal overview of non-linear time series analysis from a chaos perspective," in *Exploration of a Nonlinear World: An Appreciation of Howell Tong's Contributions to Statistics*, World Scientific, 2009, pp. 183–229.

[7] S. B. Jadhav, V. R. Udupi, and S. B. Patil, "Identification of plant diseases using convolutional neural networks," *Int. J. Inf. Technol.*, vol. 13, pp. 1–10, 2020.

[8] M. C. Golumbic, "Research issues and trends in spatial and temporal granularities," *Ann. Math. Artif. Intell.*, vol. 36, no. 1, pp. 1–4, 2002, doi:10.1023/A:1015869832085

[9] G. K. Kang, J. Z. Gao, S. Chiao, S. Lu, and G. Xie, "Air quality prediction: Big data and machine learning approaches," *Int. J. Environ. Sci. Dev.*, vol. 9, no. 1, pp. 8–16, 2018.

[10] D. Liang and N. Kumar, "Time-space Kriging to address the spatiotemporal misalignment in the large datasets," *Atmos. Environ.*, vol. 72, pp. 60–69, 2013, doi:10.1016/j.atmosenv.2013.02.034

[11] J. L. Pearce, S. L. Rathbun, M. Aguilar-Villalobos, and L. P. Naeher, "Characterizing the spatiotemporal variability of $PM_{2.5}$ in Cusco, Peru using kriging with external drift," *Atmos. Environ.*, vol. 43, no. 12, pp. 2060–2069, 2009, doi:10.1016/j.atmosenv.2008.10.060

[12] S. K. Sahu and K. V. Mardia, "A Bayesian kriged Kalman model for short-term forecasting of air pollution levels," *J. R. Stat. Soc. Ser. C Appl. Stat.*, vol. 54, no. 1, pp. 223–244, 2005, doi:10.1111/j.1467-9876.2005.00480.x

[13] Z. Ghaemi, A. Alimohammadi, and M. Farnaghi, "LaSVM-based big data learning system for dynamic prediction of air pollution in Tehran," *Environ. Monit. Assess.*, vol. 190, no. 5, 2018, doi:10.1007/s10661-018-6659-6

[14] P. Gupta and S. A. Christopher, "Particulate matter air quality assessment using integrated surface, satellite, and meteorological products: Multiple regression approach," *J. Geophys. Res. Atmos.*, vol. 114, no. D14, Jul. 2009, doi: 10.1029/2008JD011496

[15] S. Abirami, P. Chitra, R. Madhumitha, and S. R. Kesavan, "Hybrid spatio-temporal deep learning framework for Particulate Matter ($PM_{2.5}$) concentration forecasting," in *2020 International Conference on Innovative Trends in Information Technology (ICITIIT)*, 2020, pp. 1–6, doi:10.1109/ICITIIT49094.2020.9071548

[16] S. Abirami and P. Chitra, "Chapter Fourteen – Energy-efficient edge based real-time healthcare support system," in *The Digital Twin Paradigm for Smarter Systems and Environments: The Industry Use Cases*, vol. 117, no. 1, P. Raj and P. B. T.-A. in C. Evangeline, Eds. Elsevier, 2020, pp. 339–368.

[17] X. Li *et al.*, "Long short-term memory neural network for air pollutant concentration predictions: Method development and evaluation," *Environ. Pollut.*, vol. 231, pp. 997–1004, 2017, doi:10.1016/j.envpol.2017.08.114

[18] J. Chen, J. Lu, J. C. Avise, J. A. DaMassa, M. J. Kleeman, and A. P. Kaduwela, “Seasonal modeling of $PM_{2.5}$ in California’s San Joaquin Valley,” *Atmos. Environ.*, vol. 92, pp. 182–190, 2014, doi:https://doi.org/10.1016/j.atmosenv.2014.04.030
[19] U. Cortès, M. Sànchez-Marrè, L. Ceccaroni, I. R-Roda, and M. Poch, “Artificial intelligence and environmental decision support systems,” *Appl. Intell.*, vol. 13, no. 1, pp. 77–91, 2000.
[20] C. Li, N. C. Hsu, and S.-C. Tsay, “A study on the potential applications of satellite data in air quality monitoring and forecasting,” *Atmos. Environ.*, vol. 45, no. 22, pp. 3663–3675, 2011, doi:https://doi.org/10.1016/j.atmosenv.2011.04.032
[21] G. E. P. Box and G. M. Jenkins, “Some recent advances in forecasting and control,” *J. R. Stat. Soc. Ser. C (Appl. Stat.)*, vol. 17, no. 2, pp. 91–109, Jun. 1968, doi:10.2307/2985674
[22] P. J. García Nieto, E. F. Combarro, J. J. del Coz Díaz, and E. Montañés, “A SVM-based regression model to study the air quality at local scale in Oviedo urban area (Northern Spain): A case study,” *Appl. Math. Comput.*, vol. 219, no. 17, pp. 8923–8937, 2013, doi:https://doi.org/10.1016/j.amc.2013.03.018
[23] S. Qin, F. Liu, C. Wang, Y. Song, and J. Qu, “Spatial-temporal analysis and projection of extreme Particulate Matter (PM_{10} and $PM_{2.5}$) levels using association rules: A case study of the Jing-Jin-Ji region, China,” *Atmos. Environ.*, vol. 120, pp. 339–350, 2015, doi:https://doi.org/10.1016/j.atmosenv.2015.09.006
[24] J. Y. Zhu, Y. Zheng, X. Yi, and V. O. K. Li, “A Gaussian Bayesian model to identify spatio-temporal causalities for air pollution based on urban big data,” in *2016 IEEE Conference on Computer Communications Workshops (INFOCOM WKSHPS)*, 2016, pp. 3–8.
[25] L. E. Sucar, J. Pérez-Brito, J. C. Ruiz-Suárez, and E. Morales, “Learning structure from data and its application to ozone prediction,” *Appl. Intell.*, vol. 7, no. 4, pp. 327–338, 1997.
[26] C.-D. Neagu, N. Avouris, E. Kalapanidas, and V. Palade, “Neural and neuro-fuzzy integration in a knowledge-based system for air quality prediction,” *Appl. Intell.*, vol. 17, no. 2, pp. 141–169, 2002.
[27] G. K. Kang, J. Z. Gao, S. Chiao, S. Lu, and G. Xie, “Air quality prediction: Big data and machine learning approaches,” *Int. J. Environ. Sci. Dev.*, vol. 9, no. 1, pp. 8–16, 2018, doi:10.18178/ijesd.2018.9.1.1066
[28] D. Zhu, C. Cai, T. Yang, and X. Zhou, “A machine learning approach for air quality prediction: Model regularization and optimization,” *Big Data Cogn. Comput.*, vol. 2, no. 1, p. 5, 2018, doi:10.3390/bdcc2010005
[29] S. Abirami and P. Chitra, “Regional air quality forecasting using spatiotemporal deep learning,” *J. Clean. Prod.*, vol. 283, p. 125341, 2021.
[30] S. Du, T. Li, Y. Yang, and S.-J. Horng, “Deep air quality forecasting using hybrid deep learning framework,” *IEEE Trans. Knowl. Data Eng.*, vol. 33, pp. 1–1, 2019, doi:10.1109/tkde.2019.2954510
[31] Q. Zhu *et al.*, “Learning temporal and spatial correlations jointly: A Unified framework for wind speed prediction,” *IEEE Trans. Sustain. Energy*, vol. 11, no. 1, pp. 509–523, 2020, doi:10.1109/TSTE.2019.2897136

[32] V. D. Le and S. K. Cha, "Real-time air pollution prediction model based on spatiotemporal big data," 2018, [Online]. Available: http://arxiv.org/abs/1805.00432.

[33] C. J. Huang and P. H. Kuo, "A deep CNN-LSTM model for Particulate Matter ($PM_{2.5}$) forecasting in smart cities," *Sensors (Switzerland)*, vol. 18, no. 7, 2018, doi:10.3390/s18072220

[34] S. Lee and J. Shin, "Hybrid model of convolutional LSTM and CNN to predict particulate matter," *Int. J. Inf. Electron. Eng.*, vol. 9, no. 1, pp. 34–38, 2019, doi:10.18178/ijiee.2019.9.1.701

[35] D. Qin, J. Yu, G. Zou, R. Yong, Q. Zhao, and B. Zhang, "A novel combined prediction scheme based on CNN and LSTM for urban $PM_{2.5}$ concentration," *IEEE Access*, vol. 7, pp. 20050–20059, 2019, doi:10.1109/ACCESS.2019.2897028

[36] U. Pak, C. Kim, U. Ryu, K. Sok, and S. Pak, "A hybrid model based on convolutional neural networks and long short-term memory for ozone concentration prediction," *Air Qual. Atmos. Heal.*, vol. 11, no. 8, pp. 883–895, 2018, doi:10.1007/s11869-018-0585-1

[37] R. Chen, X. Wang, W. Zhang, X. Zhu, A. Li, and C. Yang, "A hybrid CNN-LSTM model for typhoon formation forecasting," *Geoinformatica*, vol. 23, no. 3, pp. 375–396, 2019, doi:10.1007/s10707-019-00355-0

[38] G. Jain, M. Sharma, and B. Agarwal, "Spam detection in social media using convolutional and long short term memory neural network," *Ann. Math. Artif. Intell.*, vol. 85, no. 1, pp. 21–44, 2019, doi:10.1007/s10472-018-9612-z

[39] H. Wang, L. Yu, S. Tian, Y. Peng, and X. Pei, "Bidirectional LSTM malicious webpages detection algorithm based on convolutional neural network and independent recurrent neural network," *Appl. Intell.*, vol. 49, no. 8, pp. 3016–3026, 2019.

[40] I. G. N. M. Jaya, Y. Andriyana, B. Tantular, Zulhanif, and B. N. Ruchjana, "Spatiotemporal dengue disease clustering by means local spatiotemporal Moran's index," *IOP Conf. Ser. Mater. Sci. Eng.*, vol. 621, no. 1, 2019, doi:10.1088/1757-899X/621/1/012017

Chapter 5

Model predictive-based control for supplies of medicines during COVID-19

Rajashree Taparia

CONTENTS

5.1 INTRODUCTION

Medicines before reaching the patients take the route from manufacturers to the distributors. Depending upon the requirements, the retail shop owners place orders to the distributors. The retailers manage the stock of the medicines based on the demand faced. Certain medicines have more demand than the others. Some are season specific, for example, medicines for viral fever, flu, and so on. Some of them are illness specific, for example, for cancer, neurological treatment, orthopedic treatment, and so on. One thing that is common for all the above-stated cases, the demand for these medicines is not known before and has uncertainty in them. Although the general trend for them can be known from the past data, which can be of short duration of few days or few months.

Model predictive control (MPC) is well known for its constraint handling and constraints on the states and/or control input are considered while deriving the control law [1–3]. The literature reveals that MPC is applied to the systems in which there is dynamics involved in supply of goods [4–6]. And it is also shown that control theory tools can be used to optimally manage them [7–10]. In the dynamics of such systems, the control input, which is replenishment order, has a constraint on it, that it can be positive or zero at any point of time, $u(k) \geq 0$. This is under the assumption that back

DOI: 10.1201/9781003230236-5

orders are not allowed as it may impose cost or the medicines may expire [11, 12]. MPC is based on a moving horizon strategy, in which at sample time k the future control sequence is calculated to optimize the predefined cost function and only the first element of the optimal sequence is applied to the system. At the next time instant, the horizon is shifted and a new optimal sequence at time $(k+1)$ is solved and the procedure is repeated over the entire horizon. In any system, when future demand of the goods is not known, the operations research community suggests many ways to forecast the demand such as taking simple averages, weighted averages, exponential smoothing, and so on. Therefore, rather than optimizing the cost function over entire horizon as done in other methods of control, it makes more sense to optimize it over the prediction horizon N_p for which the demand is predicted and the procedure is repeated over the shifted window [13].

5.2 SYSTEM MODEL

Before actually discussing the model predictive-based control of supply of medicines, it is essential to understand the system model for such a system. It is assumed that the stock of a particular medicine at the distributor or retailer is reviewed periodically in every review period and anticipating the future demand, orders for replenishment are issued. The general parameters for such a system are the on-hand stock of the medicine and the arriving orders. There is a demand from the patients/customers and depending upon the availability of the medicine, the supply is accomplished. One more factor associated with such an inventory system of medicines is the 'Lead Time'. It is defined as the time required to receive the medicines after the orders for their replenishment are issued. The replenishment orders affect the system dynamics after this time. It can be in hours, days, weeks, etc. The control design has to consider this latency in control. For simplicity of understanding, we consider a system whose dynamics involve a single medicine. This medicine is replenished from a single supplier with lead time L, which is an integer. This means that the orders issued for it now are received after L time period. This system is faced by an unknown, bounded, time-varying demand for this particular medicine. The stock replenishment orders $u(k)$ are issued at regular intervals on the basis of the on-hand stock, that is, the current stock level $y(k)$, the history of previous orders. The demand is modeled as an unknown bounded function of time $d(k)$ such that,

$$0 \leq d(k) \leq d_{\max}. \tag{5.1}$$

Such a definition of demand is quite general and accounts for any standard demand faced. If a sufficient number is available to satisfy the current demand, then actually met demand will be equal to the requested one. Any extra demand is the shortage of medicine at that particular point of time. The considered discrete-time system can be described in the state space as,

$$x(k+1) = Ax(k) + bu(k) + ws(k)$$

$$y(k) = q^T x(k) \tag{5.2}$$

where $x(k) = \begin{bmatrix} x_1(k) & x_2(k) & \dots & x_n(k) \end{bmatrix}^T$ is the state vector with $x_1(k) = y(k)$ representing the on-hand stock in period k and $x_i(k)$ for $i = 2 \cdots n$ are the arriving/pending orders issued. $u(k)$ and $s(k)$ represent the replenishment orders issued for the medicine and the supply of medicine from the distributor, respectively. The order of such a system is $n = (L + 1)$. A is $n \times n$ system matrix; b, wb and q are $n \times 1$ known vectors.

$$A = \begin{bmatrix} 1 & 1 & 0 & \dots & 0 \\ 0 & 0 & 2 & \dots & 0 \\ \vdots & \vdots & \vdots & \ddots & \vdots \\ 0 & 0 & 0 & \dots & 1 \\ 0 & 0 & 0 & \dots & 0 \end{bmatrix} b = \begin{bmatrix} 0 \\ 0 \\ \vdots \\ 0 \\ 1 \end{bmatrix} w = \begin{bmatrix} -1 \\ 0 \\ \vdots \\ 0 \\ 0 \end{bmatrix} q = \begin{bmatrix} 1 \\ 0 \\ \vdots \\ 0 \\ 0 \end{bmatrix} \tag{5.3}$$

5.3 PROBLEM FORMULATION

To discuss the model predictive-based control of supplies of medicines, we consider the basic model of system which has a single medicine supplied by a single supplier represented by Equation (5.2). From this model, the dynamics of the on-hand stock of the medicine, $x_1(k)$, is given as,

$$x_1(k+1) = x_1(k) + x_2(k) - s(k) \tag{5.4}$$

where $x_1(k)$ is the on-hand stock of the medicine, $x_2(k)$ is the arriving order after L time period, and $s(k)$ is the supply of the medicine at instant k. In the matrix form, this dynamics of on-hand stock of the medicine for future Np days can be written as,

$$\underset{\rightarrow(k+1)}{x_1} = \begin{bmatrix} x_1(k+1) \\ x_1(k+2) \\ \vdots \\ x_1(k+N_p) \end{bmatrix} = \begin{bmatrix} x_1(k) + x_2(k) - s(k) \\ x_1(k+1) + x_2(k+1) - s(k+1) \\ \vdots \\ x_1(k+N_p-1) + x_2(k+N_p-1) - s(k+N_p-1) \end{bmatrix} \tag{5.5}$$

where $\underset{\rightarrow(k+1)}{x_1}$ represents the prediction vector of on-hand stock of the medicine over N_p days as obtained from the state space model. The arriving order on the first day can be expressed in terms of the state of the system as,

$$x_2(k) = u(k - L) \tag{5.6}$$

or in general,

$$x_i(k+1) = u(k + i - (L+1)) \tag{5.7}$$

Therefore, the on-hand stock vector for N_p future days, in terms of arriving orders, becomes,

$$\underset{\rightarrow(k+1)}{x_1} = \begin{bmatrix} x_1(k) + u(k-L) - s(k) \\ x_1(k+1) + u(k-L+1) - s(k+1) \\ \vdots \\ x_1(k+N_p-1) + u(k+N_p-L-1) - s(k+N_p-1) \end{bmatrix} \tag{5.8}$$

The prediction vector can be partitioned in terms of on-hand stocks, replenishment orders and supplies as,

$$\underset{\rightarrow(k+1)}{x_1} = \begin{bmatrix} x_1(k) \\ x_1(k+1) \\ \vdots \\ x_1(k+N_p-1) \end{bmatrix} + \begin{bmatrix} u(k+1) \\ u(k-L+1) \\ \vdots \\ u(k+N_p-L-1) \end{bmatrix} - \begin{bmatrix} s(k) \\ s(k+1) \\ \vdots \\ s(k+N_p-1) \end{bmatrix} \tag{5.9}$$

Thus, the vector of on-hand stock of future N_p days will have current state $x_1(k)$, $s(k)$, the past inputs, future inputs and future sale terms in it.

5.4 SYSTEM DYNAMICS

Mathematically, if demand for the medicine is completely met, it means $d(k) - s(k) = 0$ where $d(k)$ and $s(k)$ represent the demand and supply for the medicine, respectively. The cost function *CF* for the medicine inventory system is formulated based on the following considerations.

1. To meet the demand to the maximum, which is directly related to customer satisfaction and also supply shortage cost as for the business is concerned.

2. Use the on-hand stock of medicine, maximum for supply, thereby reducing the holding cost of the medicine.

The cost function for the medicine inventory system is therefore expressed by the following equation:

$$CF = \sum_{j=L+2}^{N_p} \left[d_{\text{pred}}(k+j) - s(k+j) + \left| x_1(k+j) - s(k+j) \right| \right] \tag{5.10}$$

where L is the lead time, N_p is the prediction horizon. $d_{\text{pred}}(k)$ represents the predicted demand and $s(k)$ the supply of the medicine. $x_1(k)$ represents the on-hand stock of the medicine. Minimizing the CF (5.10) will account for the above-stated considerations. As we know, the first state in the state space model is the on-hand stock $x_1(k)$ and $x_2(k)\cdots x_n(k)$ are the arriving orders with $x_n(k+1) = u(k)$. Consider Equation (5.4)

$$x_1(k+1) = x_1(k) + x_2(k) - s(k) \tag{5.11}$$

where $x_1(k)$ is the on-hand stock, $x_2(k)$ is the arriving order today and $s(k)$ is the supply of the medicine. Since lead time is assumed to be L days, the orders issued for medicine today arrive after L days and are available for sale on $(L+1)^{th}$ day. The medicine which distributor can supply from his stock at any instant k will be the minimum of the demand and on-hand stock at that instant, hence $s(k) = \min(x_1(k), d(k))$. Therefore,

$$x_1(k+1) = x_1(k) + x_2(k) - \min(x_1(k), d(k)) \tag{5.12}$$

The arriving orders can be expressed in terms of the states of the system as follows:

$$x_i(k+1) = u(k+i-(L+1)) \tag{5.13}$$

Therefore if the lead time L is known, then the states of the system can be written in terms of the replenishment orders issued. To derive the control law based on MPC, the demand is predicted for a future horizon called 'prediction horizon', N_p from the previous demand data by method of averages. The predicted demand will be a sequence $d_{\text{pred}}(j)\cdots,\ \ j = (k+1),\cdots(k+N_p)$. The prediction horizon N_p is taken greater than L, that is, $N_p > L$, for obvious reasons. From the predicted future demand, the on-hand stocks for the future prediction horizon can be calculated, because today's supply and arriving order is going to affect tomorrow's on-hand stock. Since there is a latency in control of L time period, the orders issued today arrive on $(L+1)^{th}$ day, and they affect the dynamics on the $(L+21)^{th}$ day. Therefore, the cost function will not have control terms till $(L+2)^{th}$ day and therefore cost function summation is taken from $(L+2)$ to N_p. The control input sequence will therefore necessarily have $\left(N_p-(L+1)\right)$ terms. This reduction in input terms will require less iterations to optimize the cost function and will reduce the computations required in optimization. The cost function is therefore modified as,

$$CF=\sum_{j=L+2}^{N_p}\left[d_{\text{pred}}(k+j)-s(k+j)\right]+\sum_{i=1}^{N_p-(L+1)}u(k+i)-\sum_{j=L+2}^{N_p}s(k+j) \tag{5.14}$$

The MPC based medicine supply system can therefore be formulated as,

Problem – MPC: minimize cost function (5.14) with the constraint $u(k)\geq 0$.

5.4.1 Simulation results

To simulate the MPC-based control law for replenishment of medicines such that the demand for them is met and also overheads kept are not much, we consider a single medicine single supplier system with lead time L = 3 days. Therefore, the order of the system $n = L + 1 = 4$. The demand for the medicine is constructed as a slowly varying function with random slope and with maximum demand $d_{\max}$ = 20 units. Prediction horizon is taken as N_p = 7 days. The demand is predicted from the previous six days demand, over this prediction horizon by method of averages. The optimization toolbox of MATLAB is used to solve this optimization problem. The modified cost function (5.14) for the system is expressed in (5.2) and (5.3), with constraint as $u\geq 0$ is minimized. The results of the simulations are plotted in Figure 5.1. The plots include the demand faced for the medicine, corresponding supply of it, the on-hand stock maintained at the distributor in Figure 5.1(a). The orders issued to replenish the stock of medicine are shown in Figure 5.1(b) and the unfulfilled demand are shown in Figure 5.1(c). It can be observed from Figure 5.1 that the supply follows the demand pattern reasonably well, while at the same time the on-hand stock level is also within bounds. The few instances when the demand is not met are mainly during the initial few instants of the simulation, wherein the shortfall can be attributed to the distributor who is stocking up from empty.

5.5 DEMAND FORECASTING

This subsection describes how weighted averages demand forecasting can be implemented. It can be an important aspect for successful management

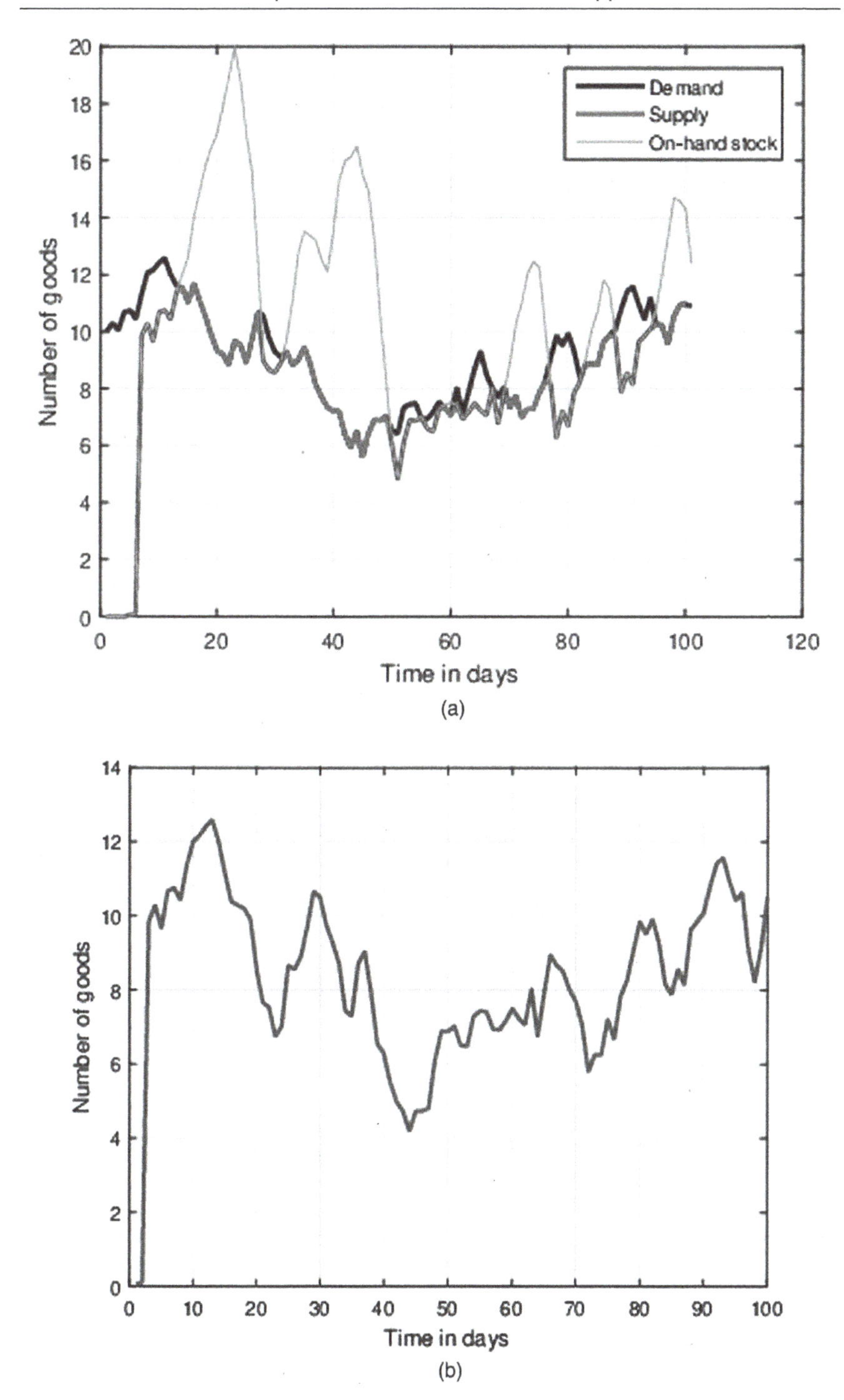

Figure 5.1 MPC-based law: replenishment of medicine.

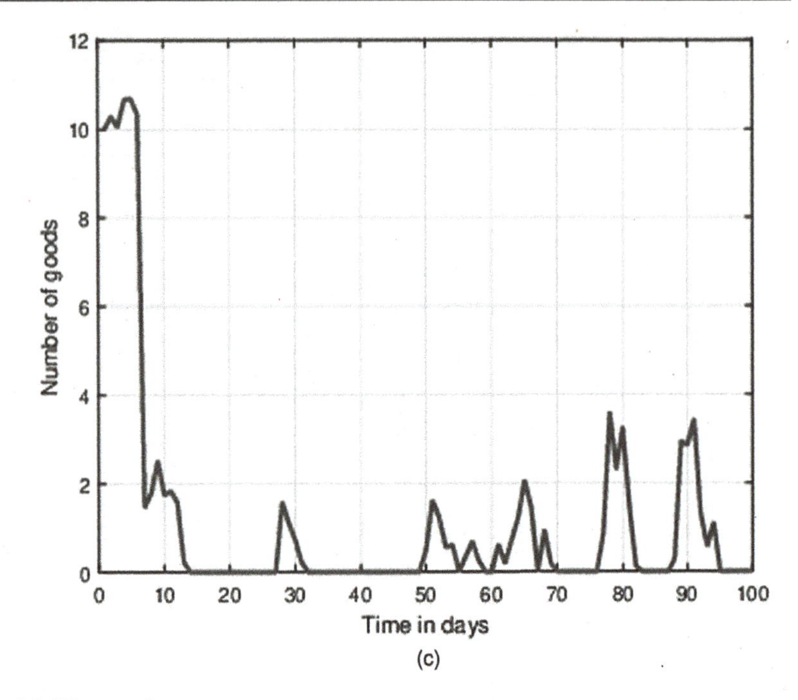

Figure 5.1 (Continued)

of the supply of medicines. The more accurate the demand prediction, the more will the orders issued be in synchronization. In operations research, many methods of demand forecasting are mentioned. In the simulations, we use a method of weighted averages for demand forecasting over the prediction horizon N_p. It is described as in Equation (5.15). The forecast with weighted averages at time t from any instant k is expressed as follows:

$$F_{(k+t)} = \frac{\sum_{i=1}^{N_p} w_i \tilde{F}_{t-i+k}}{\sum_{i=1}^{N_p} w_i} \tag{5.15}$$

where

$$\tilde{F}_{t-1+k} = \begin{cases} d_{t-i+k}, & t-i \leq 0 \\ F_{t-i+k}, & t-i > 0 \end{cases}$$

where $\tilde{F}$ denote the demand forecast. For time instants $t - i \leq 0$ all the demands are known and hence the demand forecast is taken as $\tilde{F}_{t-i+k} = d_{t-i+k}$ and for the instants $t - i > 0$, the demand forecast is $\tilde{F}_{t-i+k} = F_{t-i+k}$. The weighing factors are chosen as $w_i = \eta^{i-1}$. $0 < \eta < 1$ being a constant. Such a type of weight assignment allows more weight on the latest demand and reduces gradually the prior ones.

5.6 CONCLUSION

This work represents model predictive-based control for systems to supply medicines from distributor to retailer. For simplicity of understanding, only one medicine is considered in the model and the simulations. Whereas this can be extended to multiple number of medicines on the similar lines as described for the single case. Some of the heuristics related to medicines are formulated as a mathematical problem, and solution is found based on MPC.

REFERENCES

1. C. E. Garcia, D. M. Prett, and M. Morari, "Model predictive control: Theory and practice: A survey," *Automatica*, vol. 25, no. 3, pp. 335–348, 1989.
2. D. Mayne, J. B. Rawlings, C. Rao, and P. O. M. Scokaert, "Constrained model predictive control: Stability and optimality," *Automatica*, vol. 36, no. 6, pp. 789–814, 2000.
3. J. B. Rawlings, "Tutorial overview of model predictive control," *IEEE Control Systems*, vol. 20, no. 3, pp. 38–52, 2000.
4. J. Richalet, A. Rault, J. Testud, and J. Papon, "Model predictive heuristic control: Applications to industrial processes," *Automatica*, vol. 14, no. 5, pp. 413–428, 1978.
5. D. Fu, C. M. Ionescu, E.-H. Aghezzaf, and R. D. Keyser, "A centralized model predictive control strategy for dynamic supply chain management," *IFAC Proceedings Volumes*, vol. 46, no. 9, pp. 1608–1613, 2013.
6. R. Taparia, S. Janardhanan, and R. Gupta, "Inventory control for nonperishable and perishable goods based on model predictive control," vol. 7, no. 4, pp. 361–373, 2019.
7. H. A. Simon, "On the application of servomechanism theory in the study of production control," *Econometrica*, vol. 20, no. 2, pp. 247–268, 1952.
8. J. H. Vassian, "Application of discrete variable servo theory to inventory control," *Journal of the Operations Research Society of America*, vol. 3, no. 3, pp. 272–282, 1955.
9. A. S. White, "Management of inventory using control theory," *International Journal of Technology Management*, vol. 17, no. 7, pp. 847–860, 1999.
10. P. Ignaciuk and A. Bartoszewicz, "LQ optimal sliding-mode supply policy for periodic-review perishable inventory systems," *Journal of the Franklin Institute*, vol. 349, no. 4, pp. 1561–1582, 2012.

11. R. Taparia, S. Janardhanan, and R. Gupta, "Management of periodically reviewed inventory systems with discrete variable structure control," in *11th International Conference on Industrial and Information Systems (ICIIS)*, 2016, pp. 49–53.
12. P. Ignaciuk and A. Bartoszewicz, "Linear–quadratic optimal control strategy for periodic-review inventory systems," *Automatica*, vol. 46, no. 12, pp. 1982–1993, 2010.
13. R. Taparia, S. Janardhanan, and R. Gupta, "Laguerre function-based model predictive control for multiple product inventory systems," *International Journal of Systems Science: Operations & Logistics*, vol. 9, no. 1, pp. 133–142, 2020. [Online]. Available: https://doi.org/10.1080/23302674.2020.1846094

Chapter 6

Role of swarm intelligence for health monitoring and related actions

Pankaj Sharma, Vinay Jain, Abhishek Jain, and Mukul Tailang

CONTENTS

DOI: 10.1201/9781003230236-6

6.1 INTRODUCTION

Every time the human species progresses in terms of innovation, health becomes a major concern. The current coronavirus onslaught, which has harmed China's economy to some extent, is an illustration of how healthcare has grown increasingly important. It is often a better option to monitor such individuals utilizing remote health monitoring technologies in places where the virus has spread. Over the previous decade, the health sector has experienced remarkable expansion, contributing significantly to revenue and job creation. So an Internet of Things (IoT)-based health monitoring system is also being proposed [1, 2]. Patients with chronic diseases should be monitored on a frequent basis to avoid life-threatening situations. Because it allows us to integrate medical devices to receive and assess patient health data in the hopes of preventing crucial events, the IoT is critical for health surveillance [3]. There are three major concerns with existing IoT-based health tracking systems. For starters, they frequently use communication at somewhat high cost networks like 3G/4G/5G. Second, they rarely deal with data privacy concerns. Third, the majority of them do not examine monitored health metrics in order to avoid potentially dangerous circumstances [4–9].

Human freedom has risen as well as their capability to interact with the outside environment thanks to the IoT. The IoT is becoming a key source of international communication with the assistance of current protocols and algorithms. It links a vast number of items to the Internet, home appliances, including wireless sensors, and electrical devices [10]. The IoT has grown in popularity over the previous decade and has made everything intrinsically linked, and it has been dubbed the next technological breakthrough. IoT applications include smart health monitoring systems [11, 12], smart homes [13], smart parking [14], smart cities [15], industrial locations [16], smart climate [17], and agricultural areas [18]. The most significant application of IoT is in healthcare management, which uses it to monitor environmental and health variables. IoT is the process of connecting computers to the Internet through the use of networks and devices [19, 20]. These interconnected elements can be utilized in health-monitoring technologies. The data is subsequently sent to remote sites via machine to machine, which includes machines for computers, machines for people, smartphones, and hand-held devices [21]. It's a straightforward, energy-efficient, scalable, considerably smarter, and interoperable method of monitoring and optimizing care for any health issue. Modern technologies now include a configurable interface [22], mental health monitoring [23], and assistant devices [24] to help people live smarter lives. The IoT is gaining popularity due to its advantages of improved accuracy, reduced cost, as well as the capacity to better forecast upcoming scenarios. Furthermore, the quick IoT transformation has been aided by improving understanding of software and apps, as well as advancements in smartphones and computer advancements, the increasing prevalence of wireless technology, and the growth of the global marketplace [25].

In an IoT network, Tamilselvi et al. [26] built a health tracking system that really can monitor a patient's substantial symptoms such as pulse rate, temperature of the body, oxygen saturation percentage, and eye movement. SpO_2, Temperature, Heartbeat, and Eye Blink sensors were employed as recording components, while an Arduino-UNO was utilized as a sensor module. However, the proposed framework was deployed, and there are no defined performance metrics for any of the patients. In the IoT setting, Acharya et al. [27] presented a healthcare tracking package. Pulse, ECG, temperature, and respiration were among the primary health measures tracked by the designed architecture.

Many nations have introduced new technologies and legislation in order to optimize the use of IoT in healthcare systems. As a result, contemporary healthcare research has become a more interesting subject to investigate. The advancement of innovations will have a profound influence on every human's life and health surveillance; it will significantly reduce healthcare costs while also improving illness prediction accuracy. In this chapter, we offer a technology platform for quality care from an economic and technological standpoint, as well as open problems in integrating IoT in the actual world medical profession.

6.2 PROPOSED TECHNOLOGY

The layout and deployment of a comprehensive patient health surveillance system is the project's main focus. The suggested system is depicted in Figure 6.1. The sensors may be placed in the patient's body/room to monitor the patient's thermal expression and heartbeat. The additional sensors may be placed at home to track the temperature and humidity of the patient's room. These sensors are linked to a control unit that determines the total value of all the sensors. These computed data are then sent to the core network via an IoT cloud. The data are then accessible by hospitals at every other place from the core network. The doctor can determine the patient's condition and take necessary actions depending on the temperature and heart rate data as well as the room sensor values.

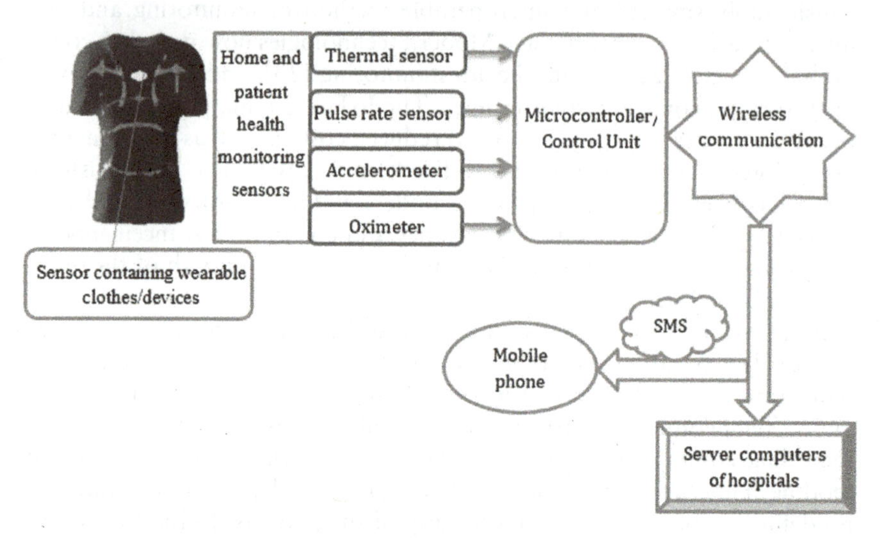

Figure 6.1 Proposed flow plan of IoT for health monitoring.

6.2.1 Sensors

Smart sensors' abilities have considerably expanded as a result of recent advancements in the IoT, mechatronics, wearable devices, and the downsizing of sensors and electronics, extending the spectrum of their applications. The development of such technologies, which are making significant contributions in a variety of applications in the area, has had a significant impact on the healthcare and medical professions, among others [28–31].

Smart sensors, in fact, are progressively providing innovative answers to a number of important issues in healthcare, like early identification of diseases and minimally invasive therapy and avoidance of high-burden disorders (e.g., cancer and cardiovascular diseases) [32]. Moreover, the advancement of tiny and lightweight smart sensor-based systems might be a major factor in accelerating the adoption of inconspicuous and unsupervised techniques to home physiotherapy and constant patient monitoring.

6.2.2 Humidity and temperature sensors

The relative humidity sensor detects relative humidity between 10% and 95%, with an average inaccuracy of 2% RH at 55% RH. The temperature sensor, on the other hand, monitors ambient temperature from –30°C to + 80°C, with an average inaccuracy of 0.75°C in the 0°C to 80°C range. The temperature sensor element is suitable for temperatures ranging from –40°C to +100°C [33]. With the assistance of an analogue-to-digital converter, the temperature sensor linked to the analogue pin of the Arduino microcontroller is transformed into a digital value [34]. The controller translates this electronic information into the real temperature value in degrees Celsius using the formula [35]:

$$\text{Temperature e}(^\circ C) = [\text{Raw ADC value} * 5 / 4095 - (400 / 1000)] * (19.5 / 1000)$$

Both humidity and air temperature are sensed, measured, and reported by a hygrometer/humidity sensor. Humidity sensors monitor changes in temperature or electrical currents in the air to determine humidity levels. The relative humidity may be determined using the formula below [35]:

$$\text{Voltage} = (\text{ADC Value}/1023) * 5$$

$$\text{Percent relative humidity} = (\text{Voltage} - 0.958) / 0.0307$$

6.2.3 Heart rate sensor

The photo plethysmography concept underpins the heartbeat sensor. It monitors the fluctuation in blood volume via any organ of the body that leads to a change in the luminance passing through that organ (a vascular region) [36]. A microcontroller receives the digital pulses and uses the algorithm to calculate the heartbeat rate by the given formula [35]:

$$\text{Heart beat rate (HBR)} = 60 * f$$

where f is the pulse frequency.

6.2.4 Accelerometer

The accelerometer is a body movement sensor that is used to anticipate the patients' motion tracking. It's a technique for keeping track of the progress of people and things. This sensor is a device that detects infrared radiation and determines if the patient should be moved to the left, right or straight posture. These sensors are specifically designed for information monitoring of the biophysical and biochemical components. The Phidget Accelerometer seems to be a three-axis that monitors three gravitational (29.4 m/s^2) changes per axis (x, y, z). This sensor can detect both dynamic and static accelerations, such as gravity or tilt [16]. The data collected from this sensor is in Table 6.1.

Table 6.1 Axis combination for determining the physical status of the patient [37]

X-axis	*Y-axis*	*Z-axis*	*Physical state of patients*
×	-	-	Sitting
×	×	+	Lying
×	×	-	Fall down

Note: × means don't care.

6.2.5 Oximeter

Ramirez Lopez et al. created an IoT architecture and used Bluetooth low-energy gadgets that conform with reduced cost and open-hardware systems that transmit and receive data from a cloud server to an edge node via "HTTP requests". To ensure the availability and authenticity of the obtained values and outputs, a network performance evaluation was carried out. SpO_2 and heart rate values were measured by the system. The most notable outcome was a 20 percent reduction in energy usage compared to competing devices [38].

6.2.6 Microcontroller/control unit

Patients will get embedded devices that can measure humidity and temperature, heart rate, body movement, oxygen level, blood glucose levels etc. Diseases including fever, arrhythmia, restlessness, weakness, low oxygen level etc. may all be treated with these devices. From these embedded sensors, physiologic data sent to the microcontroller and the acquired data are preprocessed by microcontroller. To transfer data, the microcontroller interfaces with software. The control unit elements in charge of data transfer must be able to accurately and securely convert patient recordings from any place to the health center.

Short-range low-power digital radio Zigbee or Bluetooth can be used for transmission. Additionally, the data collected can be transmitted to a health center through the Internet for archiving. The concentrator, which may even be a smartphone, can control the sensors in the IoT system through the Internet [39].

6.2.7 Wireless communication

Connectivity technologies allow diverse elements in a health-related IoT network to communicate with one another. These technologies may be classified into two categories: small- and mid-range communication. Small-range technological advancements in communication are procedures that are being used to connect objects within a narrow range or in a personal area network, so although mid-range communication technologies typically improve communications over a long distance, such as communication between a server and a personal area network's central node. In the case of small-range communication, the communication distance might range from just few millimeters to a few meters. Small-range communication technology is suitable in most healthcare IoT scenarios. The current wireless sensor communication in the health monitoring must be modified to remodify the sensing functions depending on the relative distance among sensors and the health clinic, as well as to gather additional physical information for a longer period of time by eliminating redundant jobs. Wi-Fi, Bluetooth, and other communication technologies are among the most frequently utilized [40, 41]. When focusing on low-energy usage, we should create threshold values to deal with emergency circumstances. Other sensors can be turned off at the same time to save battery life.

6.2.8 Bluetooth

When energy usage is restricted, low-power communication methods become more important. Bluetooth is a low-power wireless communication system that employs ultra-high frequency radio waves to communicate over short distances. It's ideal for a variety of purposes, including health monitoring, sports, and home entertainment. The components may be placed to sleep for lengthy periods of time using minimal energy, resulting in a significant reduction in the number of bytes delivered per joule of energy [42]. This technology enables medical equipment to communicate wirelessly with one another. Bluetooth's frequency range is 2.4 GHz. Bluetooth offers a transmission distance of up to 100 meters. Bluetooth provides data security through encryption and authentication. Bluetooth's low cost and energy efficiency are its main advantages. During data transfer, it also guarantees that there is less interruption among the linked devices. However, when a healthcare application needs long-distance communication, this approach falls short.

6.2.9 Wireless fidelity (Wi-Fi)

The IEEE 802.11 features define Wi-Fi (wireless fidelity), which is a wireless local area network. When contrasted to Bluetooth, it has a greater communication range (within 70 feet). Wi-Fi allows you to set up a network fast and simply. As a result, it is mostly utilized in hospitals. Wi-widespread Fi's use is due to its simple interoperability with cellphones as well as its ability to offer sophisticated control and security. However, it consumes significantly more power, and the network works improperly.

Smartphones nowadays are equipped with far more complex features, allowing them to function as both LTE and Wi-Fi. In this method, smartphones may function as concentrators. The concentrator's output will be sent to the cloud for storage. If this information is saved, it will be very useful for clinicians to retrieve on demand or for analytics. When local resources are insufficient to meet the needs, a tiny control system termed cloudlet is utilized for both processing and storing locally. It also aids in the execution of time-critical activities on the medical data of patients. When data is kept in a cloudlet, it is accessible at all times, allowing data analytics to generate more accurate diagnostic information.

Because health applications frequently cope with offline data, cloudlet computing has been recommended as a better option for them using personal art network. To decrease data transmission delay for important operations on the gathered data, the concentrator and cloudlet are permitted to interact through Wi-Fi interface. Finally, the cloudlet's information will be kept in the cloud for secure storage and dispersed access. Context aware concentration distinguishes the data aggregation conducted by cloud and cloudlet, where context refers to the patient's present and predicted state. Maintaining the security of a patient's electronic health records while keeping them on the cloud is becoming critical. When transferring offline data to the cloud, suitable privacy-preserving procedures should be applied to avoid unwanted access. As a result, encrypted cloud methodologies have been proposed to cope with delicate medical data, but it remains a difficulty [43].

6.2.10 IoT server

Sensors detect the patient's physiological indications whenever he or she enters the treatment center's premises, and these signs are converted to electrical signals [44]. Then, the basic electrical flag is upgraded to an upgraded flag (computerized data) and stored in radio-frequency identification (RFID). The Zigbee Protocol is used to transfer computerized data to a local server. Zigbee is an acceptable standard for this architecture. There are most cell hubs in this area. It is preferred for gadgets that are smaller in size and use less energy. Data is sent to the therapeutic server via a wireless local area network from a nearby server. When the data is sent to the therapeutic server, it verifies to see if the patient has a previous medical record; therefore, the server integrates the new data to that record and sends it to the specialist. If a patient has no previous therapy records, the server generates a new ID and keeps the data in its database [45].

6.2.11 Notifications of an emergency

This system delivers email and SMS notifications to the program's intended users (health professionals/doctors) when a sensor value exceeds a predetermined limit or threshold value, just so the individual is conscious of their health and may take appropriate action as shown in Table 6.2. In the event of an emergency, such as when sensor data exceed the threshold levels,

Table 6.2 Various Internet of Things (IoT) technologies for application in healthcare

Technologies	*Description*
Big data	• Today, big data appears to be the ideal approach for capturing, storing, and analyzing patient's information. • Data was typically saved in hard form in the medical field, which added to the expense. • Big data has a great deal of digital data storage capabilities. • All patient data, payment system, and medical record are well maintained using big data. • Data is organized in a way that rapidly delivers the greatest healthcare answer [46–49].
Cloud computing	• It saves on-demand storage space and analyzes data utilizing computer network resources and the Internet. • In emergency situations, it swiftly communicates patient information. • Share confidential data resources that assist doctors and surgeons in doing their duties more efficiently and successfully. • It improves data quality while lowering data storage costs [50–53].
Smart sensors	• In the medical field, smart sensors have a strong capacity to communicate via a digital network and generate accurate and consistent results. • It keeps track of and regulates all aspects of the patient's health. • Easily keep track of the patient's blood pressure, temperature, infusion, sugar level, oxygen concentrator, and fluid management system. • Acquiring access to healthcare condition, faulty bone, and accompanying biological tissue has never been easier [54–56].
Software	• There is specialized software available for the medical profession that helps to enhance patient care, preserve patient data, and diagnose patients. • It aids in the improvement of patient–doctor communication. • The patient's medical records, personal information, and disease are all readily detected and maintained via software [57–59].
Artificial intelligence	• In a predetermined context, artificial intelligence aids in the performance, evaluation, validation, prediction, and analysis of data. • Provide strong skills for predicting and controlling microbial infections. • Doctors and surgeons benefit from this technology because it improves their efficiency, precision, and efficacy. • It assesses the patient's discomfort as a result of drug adjustments [60–62].
Actuators	• In a particular context, an actuator is a process that provides motion and regulates the system to act upon. • Medical actuators are mostly used to ensure precision and regulate essential parameters. • Assists in the development of a hospital bed that may be raised or lowered to meet the needs of the patients [63–64].
Augmented reality/ virtual reality	• A better approach to connect human and technological technologies in order to deliver real-time data. • Virtual reality aids in to enhance the effectiveness of planning, quality care, and treatment efficiency. • To enhance surgery planning quality, provide necessary information to patients and clinicians [65–67].

the practitioner will receive an alarm notification, from which male/female will be ready to comprehend about the person's health state and make additional decisions. If the situation is urgent, he or she will provide the patient a warning and diet advice.

6.3 OBSTACLES AND CONSTRAINTS

In the last few years, the healthcare profession has seen notable technical innovations and their use in the intentions of healthcare-related challenges. This has considerably enhanced quality healthcare, which has become available at the push of the button. IoT has effectively transformed the healthcare business through the use of cloud computing, smart sensors, and modern communications. IoT, like other innovations, has its own set of difficulties and issues that might be explored more in the future. In the next part, we'll go through some of the difficulties.

6.3.1 Expenses for servicing and upkeep

Rapid technical advances have occurred in recent years, necessitating periodic upgrades of health-related IoT devices. A huge number of linked medical devices and sensors are used in every IoT-based system. This entails significant maintenance, service, and upgrade expenses, which may have an influence on the company's as well as end users' financials. As a result, sensors that can be maintained with fewer maintenance costs must be included.

6.3.2 Consumption of energy

The majority of IoT devices are powered by batteries. It is difficult to change a sensor's battery once it has been installed. As a result, a high-capacity battery was utilized to power the system. Currently, innovators throughout the world are attempting to create healthcare equipment that can produce their own electricity. Integration of the IoT device with renewable sources is one such decent choice. To some extent, these methods can assist in reducing the global energy problem.

6.3.3 Confidentiality and security of data

The notion of true tracking has been swapped by the amalgamation of cloud computing. Although, this has led the susceptibility of healthcare systems towards cyberattacks. This might result in misapply of critical patient information and have an impact on the treatment method. Several preventative steps must be considered while building a system to protect an IoT system from this harmful assault. Identity validation, safe booting, authorization management, fault tolerance, whitelisting, password protection, and safe coupling protocols must all be evaluated and used by medical and sensory equipment that are part of an IoT network to prevent an intrusion. Bluetooth, Wi-Fi, Zigbee, and other network protocols, for example, must be linked with secure routing methods and message authentication procedures. Because the IoT is a linked network in which each client is linked to a server, any flaw in the IoT protection services might jeopardize the patient's privacy. This might be addressed by incorporating sophisticated and secured cryptographies and algorithms into a more protected manner.

6.3.4 The administration of data

Because data management is so important in the IoT world, linked items create and share vast amounts of data that must be privileged mode and handled. It would be preferable if some businesses provided sufficient storage to handle all IoT data.

6.3.5 Excavation of new diseases

With the fast advancement of mobile innovation, new healthcare applications are being introduced on a daily basis. Despite the fact that a significant number of mobile apps for healthcare are accessible, the ailments for which these programs were created are still restricted. As a result, there is a need to cover additional illnesses that were previously ignored or given insufficient attention. This will increase the variety of IoT applications.

6.4 POTENTIAL FUTURE APPLICATIONS

IoT will be used in the future to watch a patient's vital signs instantaneously. This system will digitally capture all comprehensive information in order to avoid future difficulties with the patient's therapy. Doctors will be required to employ the latest technology, which will result in a significant improvement in healthcare practice. IoT is a complex emerging technology with a wide range of applications in giving actual medical treatment, allowing for a more efficient analysis of important data, information, and examination. The future has applications in medical inventory management and the healthcare supply chain to ensure that the appropriate item is delivered at the right time and location. The smart gadget of the IoT would function independently. Data will be stored on the cloud, both personal and public, and software will be available as well, making illness detection and follow-up more efficient. In the Medical 4.0 context, this revolutionary information system breakthrough will allow smarter healthcare service (Figure 6.2).

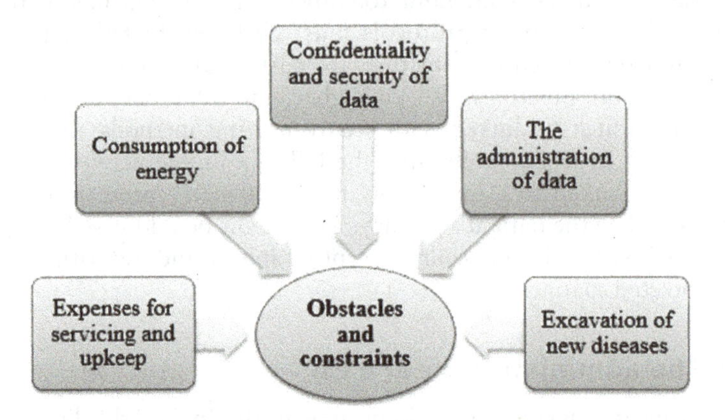

Figure 6.2 Classical applications of health-related IoT services.

6.4.1 Health-related Internet of Things services

By giving answers to numerous healthcare challenges, services and ideas have revolutionized the healthcare sector. With rising healthcare needs and technological advancements, additional services are being provided on a daily basis. These are increasingly becoming an important component of the process of developing a health-related IoT platform. In a health-related IoT ecosystem, each service delivers a collection of health technology. These ideas/services are not defined in a special manner. The applications are what make health-related IoT systems stand out. As a result, defining any idea in a broad sense is difficult. However, in order to provide context, the next section describes several of the most commonly utilized IoT healthcare services (Figure 6.3).

6.4.2 Ambient assisted living (ALL)

AAL is a subset of artificial intelligence (AI) that works in conjunction with the IoT to help the elderly. The major goal of AAL is to assist older persons in living independently in the community in a safe and convenient manner. AAL offers a method for monitoring elderly patients in real time and ensuring that they get human service-like support in the event of a medical emergency. Advanced AI systems, machine learning, big data analytics, and related use in the healthcare industry make this feasible. In general, the investigators have looked at three core fields of AAL: environment identification, activity detection, and essential surveillance. However, object detection drew the greatest attention since it relates with identifying possible dangers or critical health problems that might harm senior people' well-being. The use of IoT in AAL has been documented in a number of researches [68–71].

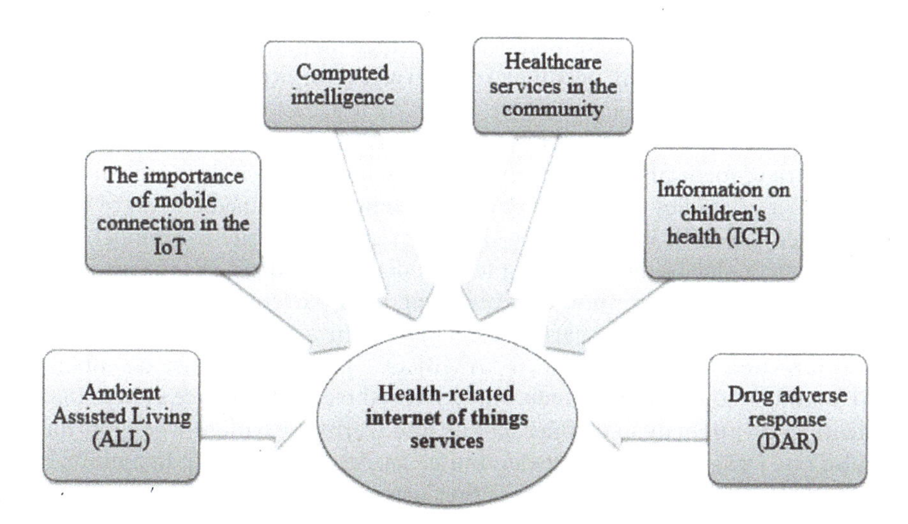

Figure 6.3 Classical applications of health-related IoT services.

Shahamabadi [72] suggested a system for delivering healthcare to the aged. For the AAL, the author created a modular design for mechanization, protection, and connectivity. IPv6-based steep-power wirelessly asynchronous transfer mode systems (6LoWPAN) [73], RFID, and NFC were utilized as communication protocols during development. To link the patient with the health professionals, the gadget uses shuttered multimedia services. The aforementioned structure was then utilized to build a more comprehensive protocol that may be employed to design complex IoT-based AAL platforms (devices, smart objects, and kits). Sandeepa recently completed a research that resulted in the development of an alarm detector for the aged that aids in the surveillance of chronic diseases and other possible health-related crises. In addition, the technology notified caretakers in the emergency situation [74]. With the aid of autonomous devices, IoT-based healthcare facilities can now check indoor air quality. These devices monitor the air quality in the patient's surroundings [75, 76] and send caregivers warnings whenever the air quality falls below a certain threshold. Gomes et al. [77] looked at how cloud computing and IoT may be used to provide a safe, accessible, and adaptable infrastructure for AAL that uses an IoT-based interface. The gateway aided in the resolution of numerous protection, data storage, and compatibility concerns in the IoT technology.

6.4.3 The importance of mobile connection in the IoT

The combination of mobile computers, sensors, communication systems, and cloud computing to monitor patient health monitoring and other physiological variables is known called m-IoT or mobile IoT. To put it another way, it acts as a communication bridge mobile networks (including 4G and 5G) and across personal area networks in order to deliver a reliable web-related healthcare service [73]. The usage of mobile has rendered health-related IoT care more affordable to medical professionals, who can now retrieve medical data, identify problems, and help people more quickly. A number of studies on the use of mobile computing in medicine have been published [78–80]. Istepanian et al. [81] created a mobile-IoT model that could detect glucose levels in people with diabetes and aid in the treatment of hypoglycemia. From another experiment, multiple sensors were employed for object tracking and heart rate regulation in a mobile way station of health-related IoT model named "AMBRO". It might also use an inbuilt GPS module to find the patients. An IoT-based true surveillance system that identifies an anomaly in cardiac activity and warns the patient whenever the pulse rate exceeds 60–100 pulses/minute was described in [82]. In a mobile-IoT model, the privacy concerns of the customer and their data are critical. Human and intellectual protections, cyber security, approval matrix, and technological guidelines are among the approaches recommended in [83] to solve the aforementioned concerns.

6.4.4 Computed intelligence

The process of analyzing difficulties in the same way that the human mind does is referred to as computed intelligence. IoT devices are now integrated with sensors which can simulate human intellect in addressing issues, thanks to recent developments in AI and sensor technology. In an IoT system, computed intelligence aids in the analysis of previously unknown patterns in huge amounts of data [84]. It also improves a sensor's capacity to handle healthcare data patterns adjust to its surroundings. These sensors in smart IoT systems function along with other smart devices to deliver effective healthcare services. The application of computed intelligence in an IoT system enables healthcare practitioners to effectively monitor medical data and give appropriate therapy. An EEG-based pervasive healthcare surveillance system that employs computed intelligence to determine the patient's clinical status has been suggested. The EEG data was combined with certain other sensor information including voice, gesture, bodily motion, and nonverbal cues to determine a patient's health. It also makes emergency assistance easier in the event of pathogenic situations [85]. Kumar et al. [86] presented a technique for transmitting intellectual information that may efficiently identify, record, and evaluate patient health information. The information of a patient in serious condition is provided with the highest priority during such event.

6.4.5 Healthcare services in the community

Community-based healthcare surveillances is the notion of establishing a healthcare system that spans surrounding residents, including a fertility hospital, a tiny housing estate, a hotel, and so forth, in order to monitor the residents' health problems. Several networks are spliced and can operate collaboratively to provide a cooperative solution in a community-based system. Aim of providing healthcare surveillance in rural regions, an IoT-based collaborative medical system was established. Various identification and authorization techniques were used to build a secure communication between the domains. A "digital clinic" was suggested as part of a community healthcare system. This aided in the provision of medical services to those in need from a faraway place. It has been suggested that a regional health system be established. A four-layer structural architecture for exchanging healthcare data, including patient medical data, was developed. This data may be accessible by health centers in order to offer correct health advice to people in the area [87, 88].

6.4.6 Information on children's health (ICH)

The idea of ICH is to raise public awareness about a child's excellently. The major goal of ICH is to inform and give direction to children and their caregivers about their entire health, encompassing nutritional content, psychological and emotional well-being, and behavior. With the construction of a system which can monitor and control a child's health, scientists were able to achieve their aim thanks to the use of IoT. Nigar and Chowdhury [89] have created an IoT-based approach to facilitate a child's physically and mentally condition. Furthermore, in the event of an emergency, appropriate actions can be performed with the assistance of physicians and parents. Another IoT-based healthcare system that links a medical equipment to a mobile app has been created. Height, SpO_2, temperature, weight, and pulse rate are among the five bodily characteristics collected by the device. The app makes this data available to physicians and other healthcare experts [90]. The ICH program offers a framework for youngsters to improve their quality of life by modifying their dietary behaviors while they are not with their parents or dining outside of their residences. Physicians, caregivers, or parents can send notifications to children who use a smart device to measure their food intake. The enhanced program can provide users with food-related data. The healthcare professional utilizes the saved data to give the kids feedback and sends a message to their caregivers. This allows parents to engage with their children and improve their family's dietary behaviors [91].

6.4.7 Drug adverse response (DAR)

An adverse effect of taking a medicine is known as a DAR. The response might happen after a single dosage or over a lengthy period of time. This can also happen as a result of an unpleasant response that occurs when two distinct drugs are taken simultaneously moment. DAR is independent of the type of drug or condition being treated, and it differs from person to person. An individual model is analyzed to recognize each drug at the patient's terminal in an IoT-based DAR network [92]. A pharmaceutical smart database may verify information regarding the medication's compliance with the person's body. Employing e-health data, the information technology saves individual sufferer's allergy profile. A judgment is made not whether the medicine is appropriate for a patient following assessing the allergy pattern and other critical health facts.

6.5 REAL-LIFE CHALLENGES AND THEIR SOLUTIONS

6.5.1 Scheduling/load balancing

In the agenda, the attention is already on the job's position relative instead of its immediate predecessor or successor, as well as the summing assessment rule/worldwide pheromone assessment rule is used.

6.5.2 Clustering

A cluster is a group of entities that are both similar and different from the entities in other clusters.

6.5.3 Optimization

The task of selecting the optimal solution/lowest cost alternative from all the possible options is known as an optimization process.

6.5.4 Routing

This is premised on the idea because rearward ants might benefit from the knowledge gained by advancements throughout their journey from transmitter to the receiver.

6.6 CONCLUSION

Increasingly emerging innovations are fast being helpful in the healthcare industry, such as gadgets and models that check vital signs on a regular basis, as well as other devices which monitor true medical data. Patients and physicians are utilizing mobile-based system to support their health needs as Internet speeds have increased and smartphones have become more widely available. The IoT is currently widely regarded as one of the most practical options for remote value assessment, particularly in the field of health surveillance. It enables the secure storage of personal health metric information in the cloud, the reduction of hospital visits for regular tests, and, most importantly, the monitoring and diagnosis of disease by any physician at any location. An IoT-based health monitoring system was created in this article. The system used sensors to measure temperature of the body, room humidity and temperature, and pulse rate which were also shown on a screen. These sensor data are subsequently wirelessly transmitted to a medical server. These details are then delivered over the IoT network to an authorized individual's mobile. The physician then diagnoses the ailment and the physician's health status based on the results. As a result, the healthcare sector has shifted from a hospital-centric to a patient-centric model. We've also spoken about the different uses of the healthcare IoT system, as well as their current trends. The problems and issues related to the design, production, and usage of the healthcare IoT system were also discussed. In the next decades, these issues will serve as a foundation for future progress and research focus. Furthermore, readers who are interested in not only starting their study but also making breakthroughs in the field of healthcare IoT gadgets will get thorough up-to-date information on the gadgets.

ACKNOWLEDGMENTS

The authors express special thanks to Suman Jain (Director, School of Studies in Pharmaceutical Sciences, Jiwaji University, Gwalior) and Navneet Garud for their kind support (Department of Pharmaceutics, School of Studies in Pharmaceutical Sciences, Jiwaji University, Gwalior).

REFERENCES

[1] Ali Z, Hossain MS, Muhammad G, Sangaiah AK (2018) An intelligent healthcare system for detection and classification to discriminate vocal fold disorders. Future Gener Comput Syst 85:19–28.

[2] Almotiri SH, Khan MA, Alghamdi MA (2016) Mobile health (m-health) system in the context of IoT. In 2016 IEEE 4th International Conference on Future Internet of Things and Cloud Workshops (FiCloudW), pp. 39–42.

[3] Mdhaffar A, Chaari T, Larbi K, Jmaiel M, Freisleben B (2017) IoT-based health monitoring via LoRaWAN. In IEEE EUROCON 2017 – 17th International Conference on Smart Technologies, pp. 519–524.

[4] Yang G, Xie L, Antysalo MM, Zhou X, Pang Z, Xu LD, Kao-Walter S, Chen Q, Zheng L (2014) A health-IoT platform based on the integration of intelligent packaging, unobtrusive bio-sensor, and intelligent medicine box. IEEE Trans Industrial Informatics 10(4):2180–2191.

[5] Rasid MF, Musa WM, Kadir NA, Noor AM, Touati F, Mehmood W, Khriji L, Al-Busaidi A, Mnaouer AB (2014) Embedded gateway services for internet of things applications in ubiquitous healthcare. In 2nd International Conference on Information and Communication Technology (ICoICT), pp. 145–148.

[6] Ahmed MU, Jorkman MB, Causevic A, Fotouhi H, Linden M (2015) An overview on the internet of things for health monitoring systems. In 2nd EAI International Conference on IoT Technologies for HealthCare, pp. 1–7.

[7] Istepanian RSH, Hu S, Philip NY, Sungoor A (2011) The potential of internet of m-health things "m-iot" for non-invasive glucose level sensing. In 2011 Annual International Conference of the IEEE Engineering in Medicine and Biology Society, pp. 5264–5266.

[8] Takpor TO, Atayero AA (2015) Integrating internet of things and ehealth solutions for students healthcare. In Proceedings of the World Congress on Engineering 1:256–268.

[9] Jara AJ, MZamora-Izquierdo MA, Gmez-Skarmeta AF (2013) Interconnection framework for mhealth and remote monitoring based on the internet of things. IEEE Journal on Selected Areas in Communications 31(9):47–65.

[10] Khan M, Han K, Karthik S (2018) Designing smart control systems based on internet of things and big data analytics. Wirel Pers Commun 99(4):1683–1697.

[11] Rahaman A, Islam M, Islam M, Sadi M, Nooruddin S (2019) Developing IoT based smart health monitoring systems: a review. Rev Intell Artif 33:435–440.

[12] Riazul ISM, Kwak D, Kabir MH, Hossain M, Kwak KS (2015) The Internet of Things for health care: a comprehensive survey. IEEE Access 3:678–708.

[13] Al-Ali AR, Zualkernan IA, Rashid M, Gupta R, Alikarar M (2017) A smart home energy management system using IoT and big data analytics approach. IEEE Trans Consum Electron. https:// doi.org/10.1109/TCE.2017.015014

[14] Lin T, Rivano H, Le Mouel F (2017) A survey of smart parking solutions. IEEE Trans Intell Transp Syst 18:3229–3253. https://doi.org/10.1109/TITS.2017.26851 43

[15] Zanella A, Bui N, Castellani A, Vangelista L, Zorzi M (2014) Internet of Things for smart cities. IEEE Internet Things J 1:22–32. https://doi.org/10.1109/JIOT.2014.23063 28

[16] Chen B, Wan J, Shu L, Li P, Mukherjee M, Yin B (2018) Smart factory of Industry 4.0: key technologies, application case, and challenges. IEEE Access 6:6505–6519. https://doi.org/10.1109/ACCES S.2017.27836 82

[17] Mois G, Folea S, Sanislav T (2017) Analysis of three IoT-based wireless sensors for environmental monitoring. IEEE Trans Instrum Meas 66:2056–2064. https://doi.org/10.1109/TIM.2017.26776 19

[18] Ayaz M, Ammad-Uddin M, Sharif Z, Mansour A, Aggoune E-HM (2019) Internet-of-Things (IoT)-based smart agriculture: toward making the fields talk. IEEE Access 7:129551–129583. https://doi.org/10.1109/ACCES S.2019.29326 09

[19] Hasan M, Islam MM, Zarif MII, Hashem MMA (2019) Attack and anomaly detection in IoT sensors in IoT sites using machine learning approaches. Internet Things 7:100059. https://doi.org/10.1016/j.iot.2019.10005 9

[20] Nooruddin S, Milon Islam M, Sharna FA (2020) An IoT based device-type invariant fall detection system. Internet Things 9:100130. https://doi.org/10.1016/j.iot.2019.10013 0
[21] Islam M, Neom N, Imtiaz M, Nooruddin S, Islam M, Islam M (2019) A review on fall detection systems using data from smartphone sensors. Ingenierie des systemes d Inf 24:569–576. https://doi.org/10.18280 / isi.24060 2.
[22] Mahmud S, Lin X, Kim J-H, Iqbal H, Rahat-Uz-Zaman M, Reza S, Rahman MA (2019) A multi-modal human machine interface for controlling a smart wheelchair. In 2019 IEEE 7th Conference on Systems, Process and Control (ICSPC). IEEE, pp. 10–13.
[23] Lin X, Mahmud S, Jones E, Shaker A, Miskinis A, Kanan S, Kim J-H (2020) Virtual reality-based musical therapy for mental health management. In 2020 10th Annual Computing and Communication Workshop and Conference (CCWC). IEEE, pp. 948–952.
[24] Mahmud S, Lin X, Kim JH (2020) Interface for human machine interaction for assistant devices: a review. In 2020 10th Annual Computing and Communication Workshop and Conference (CCWC). IEEE, pp. 768–773.
[25] Jagadeeswari V (2018) A study on medical Internet of things and Big Data in personalized healthcare system. Health Information Science and Systems 6:14.
[26] Tamilselvi V, Sribalaji S, Vigneshwaran P, Vinu P, GeethaRamani J (2020) IoT based health monitoring system. In 2020 6th International Conference on Advanced Computing and Communication Systems (ICACCS). IEEE, pp. 386–389.
[27] Acharya AD, Patil SN (2020) IoT based health care monitoring kit. In 2020 Fourth International Conference on Computing Methodologies and Communication (ICCMC). IEEE, pp. 363–368.
[28] Bellagente P, Bona M, Gorni D (2015) The use of smart sensors in healthcare applications. Review Appl Mech Mater 783:29–41. doi:10.4028/www.scientific.net/AMM.783.29
[29] Pramanik PKD, Upadhyaya BK, Pal S, Pal T (2019) Internet of things, smart sensors, and pervasive systems: Enabling connected and pervasive healthcare. Healthcare data analytics and management. Cambridge, MA: Academic Press, pp. 1–58.
[30] Kim J, Campbell AS, De Avila BE-F, Wang J (2019) Wearable biosensors for healthcare monitoring. Nat Biotechnol 37:389–406. doi:10.1038/s41587-019-0045-y
[31] Kenry, Yeo JC, Lim CT (2016) Emerging flexible and wearable physical sensing platforms for healthcare and biomedical applications. Microsyst Nanoeng 2:16043. doi:10.1038/micronano.2016.43
[32] Andreu-Perez J, Leff DR, Ip HMD, Yang GZ (2015) From wearable sensors to smart implants – toward pervasive and personalized healthcare. IEEE Trans Biomed Eng 62:2750–2762.
[33] Phidgets Inc. (2015) Available: [Online] www.phidgets.com, last accessed: 20 October 2015.
[34] Rahman RAb, Abdul-Aziz NS, Kassim M, Yusof MI (2017) IoT-based personal health care monitoring device for diabetic patients. IEEE 978(1):5090–5152.

[35] Valsalan P, Ahmed T, Baomar B, Hussain A, Baabood O (2020) IOT based health monitoring system. J Crit Rev 7(4):20.
[36] Darshan KR, Anandakumar KR (2015) A comprehensive review on usage of internet of things (IoT) in healthcare system. In Proceedings of the International Conference on Emerging Research in Electronics, Computer Science and Technology, 2015.
[37] Ghosh AM, Halder D, Alamgir-Hossain SK (2016) Remote health monitoring system through IoT, 2016 5th International Conference on Informatics, Electronics and Vision (ICIEV).
[38] Lopez RLJ, Aponte PG, Garcia RA (2019). Internet of things applied in healthcare based on open hardware with low-energy consumption. Healthc Inform Res 25(3):230–235. https://doi.org/10.4258/hir.2019.25.3.230
[39] Sathya M, Madhan S, Jayanthi K (2018) Internet of things (IoT) based health monitoring system and challenges. Int J Eng Technol 7(1.7):175–178.
[40] Pradhan B, Bhattacharyya S, Pal K (2021) IoT-based applications in healthcare devices. J Healthc Eng 19. www.hindawi.com/journals/jhe/2021/6632599/
[41] Chou CT, Rana R, Hu W (2009) Energy efficient information collection in wireless sensor networks using adaptive compressive sensing. In IEEE 34th Conference on Local Computer Networks, LCN, pp. 443–450.
[42] Siekkinen M, Hiienkari M, Nurminen J, Nieminen J (2012) How low energy is bluetooth low energy? comparative measurements with zigbee/802.15.4. In Wireless Communications and Networking Conference Workshops (WCNCW). IEEE, pp. 232–237.
[43] Li M, Yu S, Zheng Y, Ren K, Lou W (2013) Scalable and secure sharing of personal health records in cloud computing using attribute-based encryption. IEEE Trans Parallel Distrib Syst 24(1):q131–143.
[44] Internet of Things (IoT): Number of Connected Devices Worldwide From 2012 to 2020 (in billions). [Online]. Available: www.statista.com/statistics/471264/iot-numberof-connected- devices-worldwide/
[45] Chavan P, More P, Thorat N, Yewale S, Dhade P (2016) ECG – Remote patient monitoring using cloud computing. IJIR 2(2):368–372.
[46] Raghupathi W, Raghupathi V (2014) Big data analytics in healthcare: promise and potential. Health Inf Sci Syst 2(1). doi:https://doi.org/10.1186/2047-2501-2-3
[47] Yang C, Huang Q, Li Z, Liu K, Hu F (2017) Big data and cloud computing: innovation opportunities and challenges. Int J Digit Earth 10(1):13–53.
[48] Pramanik MI, Lau RYK, Demirkan H, Azad MAK (2017) Smart health: big data-enabled health paradigm within smart cities. Expert Syst Appl 87:370–383.
[49] Jagadeeswari V, Subramaniyaswamy V, Logesh R, Kumar V (2018) A study on medical Internet of things and big data in personalised healthcare system. Health Inf Sci Syst 6(1):14. https://doi.org/10.1007/s13755-018-0049-x
[50] Sun E, Zhang X, Li Z (2012) The Internet of Things (IoT) and cloud computing (CC) based tailings dam monitoring and pre-alarm system in mines. Saf Sci 50:811–815.

[51] Sandhu R, Gill HK, Sood SK (2016) Smart monitoring and controlling of pandemic influenza A (H1N1) using social network analysis and cloud computing. J Comput Sci 12:11–22.
[52] Ferrari P, Flammini A, Rinaldi S, Sisinni E, Maffei D, Malara M (2018) Impact of quality of service on cloud-based industrial IoT applications with OPC UA. Electronics 7: 109. https://doi.org/10.3390/electronics7070109
[53] Stergiou C, Psannis KE, Kim BG, Gupta B (2018) Secure integration of IoT and cloud computing. Future Generat Comput Syst 78:964–975.
[54] Liu Z, Li Q, Yan J, Tang Q (2007) A novel hyperspectral medical sensor for tongue diagnosis. Sens Rev 27(1):57–60.
[55] Bloss R (2017) Multi-technology sensors are being developed for medical, manufacturing, personal health and other applications not previously possible with historic single technology sensors. Sens Rev 37(4):385–389.
[56] Haleem A, Javaid M (2018) Industry 5.0 and its applications in orthopaedics. J Clin Orthop Trauma 10(4):807.
[57] Zhu XJ, Tan XR, Lu N, Chen SX, Chen XJ (2016) Software solution of medical grey relational method based on SAS environment. Grey Syst Theor Appl 6(3):309–321.
[58] Carroll N, Richardson I (2016) Software-as-a-medical device: demystifying connected health regulations. J Syst Inf Technol 18(2):186–215.
[59] Akpan IJ, Udoh EA, Adebisi B (2020) Small business awareness and adoption of state-of-the-art technologies in emerging and developing markets, and lessons from the COVID-19 Pandemic. J Small Bus Enterpren 25:1–8.
[60] Pomprapa A, Muanghong D, Keony M, Leonhardt S, Pickerodt P, Tjarks O, Schwaiberger D, Lachmann B (2015) Artificial intelligence for closed-loop ventilation therapy with hemodynamic control using the open lung concept. Int J Intell Comput Cybernet 28(1):50–68.
[61] Upadhyay AK, Khandelwal K (2019) Artificial intelligence-based training learning from application. Dev Learn Org Int J 33(2):20–23.
[62] Ndiaye M, Oyewobi SS, Abu-Mahfouz AM, Hancke GP, Kurien AM, Djouani K (2020) IoT in the wake of COVID-19: a survey on contributions, challenges and evolution. IEEE Access 12(8):186821–186839.
[63] Asua E, Etxebarria V, Garcia A, Feuchtwange J (2009) Micropositioning control of smart shape-memory alloy-based actuators. Assemb Autom 29(3):272–278.
[64] Lu H, Yao Y, Lin L (2015) Temperature sensing and actuating capabilities of polymeric shape memory composite containing thermochromic particles. Pigment Resin Technol 44(4):224–231.
[65] Aksoy E (2019) Comparing the effects on learning outcomes of tablet-based and virtual reality-based serious gaming modules for basic life support training: randomized trial. JMIR Serious Games 7(2):e13442. https://doi.org/10.2196/13442
[66] Khan R, Plahouras J, Johnston BC, Scaffidi MA, Grover SC, Walsh CM (2019) Virtual reality simulation training in endoscopy: a Cochrane review and meta-analysis. Endoscopy 51(7):653–664. https://doi.org/10.1055/a-0894-4400.

[67] Janeh O, Fründt O, Schonwald B, Gulberti A, Buhmann C, Gerloff C, Steinicke F, Pötter-Nerger M (2019) Gait training in virtual reality: short-term effects of different virtual manipulation techniques in Parkinson's disease. Cells 8(5). https://doi.org/10.3390/cells8050419
[68] Marques G, Pitarma R (2016) An indoor monitoring system for ambient assisted living based on internet of things architecture. Int J Environ Res Public Health 13(11):1152.
[69] Dohr A (2010) The internet of things for ambient assisted living. In Proceedings of the 2010 Seventh International Conference on Information Technology: New Generations, Las Vegas, NA, USA, pp. 804–809.
[70] Tsirmpas C, Anastasiou A, Bountris P, Koutsouris D (2015) A new method for profile generation in an internet of things environment: an application in ambient-assisted living. IEEE Internet Things J 2(6):471–478.
[71] Maskeliunas R (2019) A review of Internet of things technologies for ambient assisted living environments. Future Internet 11:259.
[72] Shahamabadi MS (2013) A network mobility solution based on 6LoWPAN hospital wireless sensor network (NEMOHWSN). In Proceedings of the 2013 Seventh International Conference on Innovative Mobile and Internet Services in Ubiquitous Computing. Taichung, Taiwan, July 2013, pp. 433–438.
[73] Tabish R (2014) A 3G/WiFi-enabled 6LoWPAN-based U-healthcare system for ubiquitous real-time monitoring and data logging. In Proceedings of the 2nd Middle East Conference on Biomedical Engineering, Doha, Qatar, February 2014, pp. 277–280.
[74] Sandeepa C (2020) An emergency situation detection system for ambient assisted living. In Proceedings of the 2020 IEEE International Conference on Communications Workshops (ICC Workshops), Anchorage, AL, USA, June 2020, pp. 1–6.
[75] Marques G, Pires IM, Miranda N, Pitarma R (2019) Air quality monitoring using assistive robots for ambient assisted living and enhanced living environments through internet of things. Electronics 8(12):1375.
[76] Marques G, Pitarma R (2019) A cost-effective air quality supervision solution for enhanced living environments through the internet of things. Electronics 8(2):170.
[77] Gomes BdTP, Muniz LCM, Silva FJSe, Rios LET, Endler M (2017) A comprehensive and scalable middleware for ambient assisted living based on cloud computing and Internet of things. Concurr Comput Concurr Comp-Pract E 29(11):e4043.
[78] Mora H, Gil D, Terol RM, Azorin J, Szymanski J (2017) An IoT-based computational framework for healthcare monitoring in mobile environments. Sensors 17(10):2302.
[79] Tyagi S (2016) A conceptual framework for IoT-based healthcare system using cloud computing. In Proceedings of the 2016 6th International Conference-Cloud System and Big Data Engineering (Confluence), Noida, India, January 2016, pp. 503–507.
[80] Nazir S (2019) Internet of things for healthcare using effects of mobile computing: a systematic literature review. Wirel Commun Mob Comput 2019:5931315.

[81] Istepanian RSH, Casiglia D, Gregory JW (2017) Mobile health (m-Health) for diabetes management. Br J Health Care Manag 23(3):102–108.
[82] Chuquimarca L (2020) Mobile IoT device for BPM monitoring people with heart problems. In Proceedings of the 2020 international conference on electrical, communication, and computer engineering (ICECCE), Istanbul, Turkey, June 2020, pp. 1–5.
[83] AlMotiri SH (2016) Mobile health (m-health) system in the context of IoT. In Proceedings of the 2016 IEEE 4th International Conference on Future Internet of things and Cloud Workshops (FiCloudW), Vienna, Austria, August 2016, pp. 39–42.
[84] Behera RK, Bala PK, Dhir A (2019) The emerging role of cognitive computing in healthcare: a systematic literature review. Int J Med Inform 129:154–166.
[85] Amin SU, Hossain MS, Muhammad G, Alhussein M, Rahman MA (2019) Cognitive smart healthcare for pathology detection and monitoring. IEEE Access 7:10745–10753.
[86] Kumar MA, Vimala R, KRA Britto (2019) A cognitive technology based healthcare monitoring system and medical data transmission. Meas 146:322–332.
[87] Kelati A (2018) Biosignal monitoring platform using Wearable IoT. In Proceedings of the 22st Conference of Open Innovations Association FRUCT, Petrozavodsk, Russia, May 2018, pp. 9–13.
[88] Wang W (2011) The internet of things for resident health information service platform research. In Proceedings of the IET International Conference on Communication Technology and Application (ICCTA 2011), Beijing, China, May 2011.
[89] Nigar N, Chowdhury L (2019) An intelligent children healthcare system by using ensemble technique. In Proceedings of International Joint Conference on Computational Intelligence, Budapest, Hungary, November 2019, pp. 137–150.
[90] Sutjiredjeki E, Basjaruddin NC, Fajrin DN, Noor F (2020) Development of NFC and IoT-enabled measurement devices for improving health care delivery of Indonesian children. J Phys Conf Ser 1450:2020.
[91] Vazquez-Briseno M, Navarro-Cota C, Nieto-Hipolito JI, Jimenez-Garcia E, Sanchez-Lopez J (2012) A proposal for using the Internet of things concept to increase children's health awareness. In Proceedings of the CONIELECOMP 2012, 22nd International Conference on Electrical Communications and Computers, Cholula, Puebla, Mexico, February 2012, pp. 168–172.
[92] Jara AJ (2010) A pharmaceutical intelligent information system to detect allergies and adverse drugs reactions based on internet of things. In Proceedings of the 2010 8th IEEE International conference on pervasive computing and communications workshops (PERCOM Workshops), Mannheim, Germany, April 2010, pp. 809–812.

Chapter 7

Real-time implementation of an implantable antenna using chicken swarm optimization for IoT-based wearable healthcare applications

M. Bhuvaneswari, S. Sasipriya, and R. Arun Chakravarthy

CONTENTS

7.1 INTRODUCTION

Wireless implantable healthcare devices give freedoms to advanced checking of the patient, documentation, and treatment and it is controlled by electromagnetic (EM) waves. Implantable devices empower smaller than expected, battery-less implants, yet subsequently, rely upon dependable power transmission through tissue. Remote implantable equipment has no embedded battery, rather gathering energy from incoming EM waves, permitting scaling down, and expanding the life expectancy of the implant. Furthermore, these devices are especially appropriate for intermittent checking in blend with embedded biosensors, which possibly should be powered while acquiring

DOI: 10.1201/9781003230236-7

a sensor perusing. In a cognitive wireless powered communication system (CWPCN), wearable medical devices play an important role in screening patients via wireless communications. The Internet of Things (IoT) signification development in connectivity has been infiltrating every home, vehicle, medical field, and workplace as smart, Internet-connected devices. However, our reliance on newly connected devices increased as its benefits and application of a maturing technology have promising features. As every appliance, light, door, piece of clothing, and other objects in home or office become potentially Internet-enabled, the IoT is poised to put significant strain on existing Internet and data center infrastructure.

IoT devices are enabled by gateways, embedded sensors and fog computing. The main constrain in the design is the antenna selection. When designing embedded sensors to achieve consistently strong operation, miniaturization, patient security, biocompatibility, lower power consumption, a narrow frequency band, and dual-band functionality should all be considered. Because it displays the overall implant behavior, the antenna in the design of a wearable sensor is difficult to choose. A multiple-input-multiple-output (MIMO) antenna for wearable applications with Internet of medical things and fog computing in wearable medical services is critical in this work. Because it works with a variety of ready transmitters and relies on the MIMO framework for successful communication. It is critical to use a MIMO framework that is suitable for IoT and fog applications which have high behavioral quality, low power consumption, and low computational complication via reconfigurable designs.

This chapter is organized into the following sections. The literature review is discussed in Section 7.2, and proposed methodology and results and performance analysis are given in Sections 7.3 and 7.4, respectively. Conclusion and future discussions are briefed in Section 7.5.

7.2 LITERATURE REVIEW

The capacity of implantable gadgets is subject to the development of EM areas at the implanting region and the effective catch of energy utilizing antennas that are implantable [1]. The interest in the use of wireless devices expands in biomedical applications. Due to having numerously profitable, for example, following and controlling the patients distantly or being used in different disorder therapies, implantable clinical gadgets have gotten impressive interest by the analysts and clinical specialists. These gadgets ought to be planned by thinking about the way that they are being put inside the human body. This reality gives that implantable clinical gadgets ought to be biocompatible, less complex, tiny, reliable, power effective, and should ensure secure correspondence [2].

IoT is the fundamental technique, which communicates amid the natural and the digital environment. It has changed from a correlation of installed computing components to a correlation of smart sensor components. IoT comprises the machine-to-machine interaction that shares details via the Internet in the absence of human communications. It is a cloud system that is utilized to gather and manage information from the components of the sensor. The healthcare-associated information sensing methodology is a development corresponding to daily healthcare examinations out of a hospital concerning a patient's home. The sick patients must be recognized and they should be observed under the guidance of a healthcare professional. The IoT equipment gathers and transfers the health data such as degree of stress, levels of glucose, weight, and electroencephalogram. Health informatics is extremely affected by huge data of IoT components. The main factor in IoT is estimating, which is arising as a reaction to minimize data overloading and the inabilities of a system [3].

In the recent situation, wearable sensors are turned progressively very suitable in the areas of healthcare investigation and applications. Presently, minimized area and prices of many such sensors are present for observing natural and sports actions, tracing, man-PC communication, restoration, and tracing of the old people for the advantage of ambient living intentions. There shall be an extensive development as it is from the IoT, and it does not need private security measures. IoT currently exposed its iPhone healthcare app that operates similar to the health kit advancement instrument to accumulate the information of health and various functions [4]. The developers of this app can take benefit from the different kinds of health detecting equipment. They collect data from both wired and wireless sensors and alternate devices. The health observing method has affirmed the encryption and saving of health details in the cloud, and app users can share these details with healthcare professionals, hospitals, and relations as required [5].

With the development in biomedical telemetry, effective instruments with small areas are needed for biomedical utilizations. Recently, implantable medical devices have attracted investigators. They are utilized in many remote patients observing utilization like glucose observing, cardiac pacemakers, and so on. For two-directional wireless communications, antennas that can be implanted are used in such devices. Such antennas possess the capacity of transferring physiological information to the receiver instrument at the doctor's side. The model of the implantable antenna comprises particular issues like bandwidth improvement, size limitations, biocompatibility, security of the patient, and detuning procedure. Investigators are progressively trying to overcome such problems [6].

Recently, on-body health observing instruments have obtained high familiarity. The continuing reduction in the size of the sensors has made the forthcoming medical component incorporated into wearable components. The broad area of utilization is integrated with components that are wearable in the area of energetic health observing, training of military and sports personalities, private safety, and computing. Physiological specifications, like the level of glucose, the pressure of the blood, and heat level, are frequently calculated from the body of the human. Calculated details are coordinated with a personal computer or any other devices for observing and saving. Hence, the antenna acts as the major significant component of the healthcare devices for assuring satisfying transmission of power from implantable wearable devices to off-body observing devices. Many model issues emerge while modeling the implantable antennas. Near-field EM radiations communicate with multiple layered tissue models and develop interior reflection and dispersion. To surmount this problem, several model methods are employed. Such methods give better separation amid the antenna and the body of the human, yet possess a greater and complicated structure that makes them complex to incorporate with wearable instruments. Broad bandwidth is also the main essential parameter to assure reliable behavior when the antenna is located on various human bodies. A substantial change in the resonating band is noticed because of disfigurement in the structure. A higher specific absorption rate also limits the radiation of power from the antenna. In addition, multiple path fading happens because of various body positions and movements. So, to improve the signal transfer quality of wearable instruments, MIMO antennas are gaining more attraction among investigators [7].

Nowadays, for healthcare utilization, an implantable antenna is employed in the absence of wired interaction. Such antennas possess maximum reputation. Interaction in-between device that is implanted inside and external functioning equipment, such as antennas that are implantable, plays a significant part. These antennas give interaction between the doctor and the patients. With the assistance of a remote observing system, wireless interaction is employed securely. The patient is not needed to visit the hospital for consultation with the doctor often [8]. MIMO performance analysis is discussed with many techniques [9]. A modified chicken swarm optimization (CSO) algorithm is used in the synthesizing linear, circular, and random antenna arrays [10]. The sensors with the controller are in the hardware interface which are used to track the input source by consuming less power [11]. This gives the idea to look after the antenna radiation using solar power [12]. A revolutionary approach is followed to track the weather monitoring location, so any emergency testing and tracking is done with the help of analog tools. Figure 7.1 depicts the fog layer in a real-time application, illuminating the importance of fog and its role with the cloud layer and end device layer.

Hence, this work proposed an implantable MIMO IoT-based antenna and chicken swarm optimization method is employed to improving the antenna capacity for its application in wearable healthcare devices.

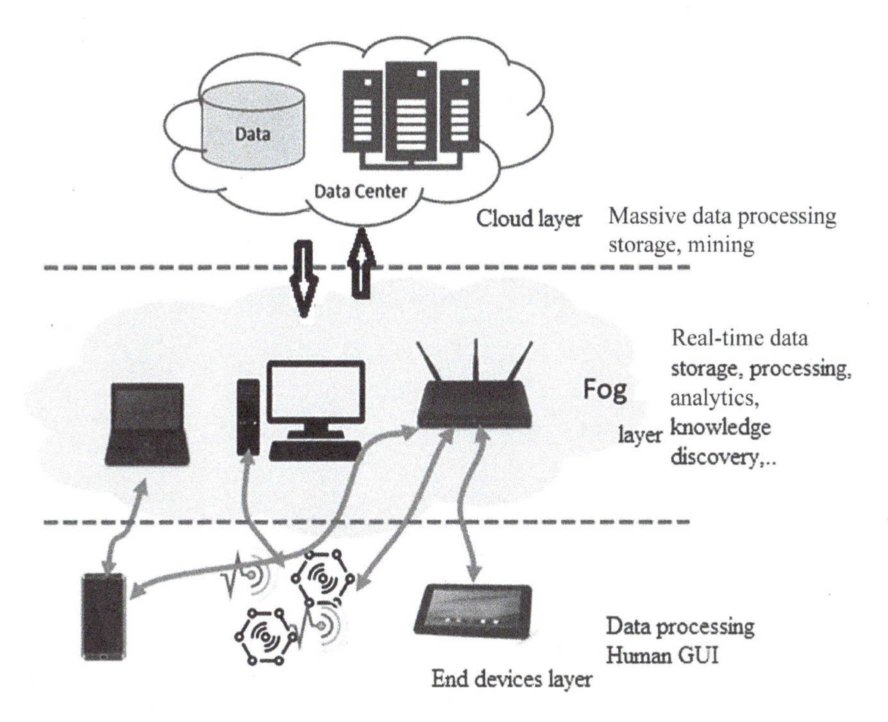

Figure 7.1 Fog layer in real-time application.

7.3 PROPOSED METHODOLOGY

As for an implantable medical device, the implantable antenna is a crucial part of interchanging information amid implantable medical devices and external base stations wirelessly. Anyhow, it also requires handling the rough radio distribution conditions, like the changes of dielectric characteristics in the tissues of the human body and the exterior disruptions obtained from human movement, extremely diminishing the behavior of implanted antennas. Furthermore, the reduction in the size of antennas must be taken into account because of the space limitations for devices that are implanted within the body of the human. So, several aspects like miniaturization, biocompatibility, security, and strong wireless interaction must be considered for the method of relevant implantable antennas in the dissimilar human tissues. Many of the implanted antennas are single-input single-output models. When its design and features are finished, area and capacity are limited. To again maximize the rate of transferring information and improve the non-dependence of locations and the immunity to multiple path interventions and mutual coupling, extra antenna features are required to be

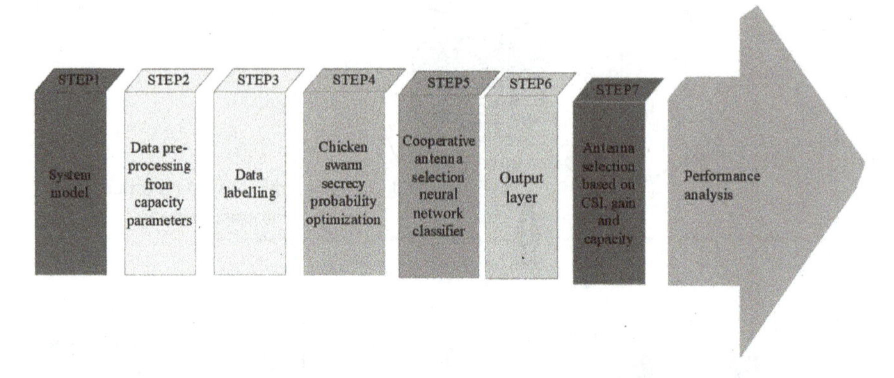

Figure 7.2 Flow of the proposed technique.

regarded and presented. The behavior of wearable antennas is adequately diminished when functioning on a regular or an irregular human body. Indeed, rigorous multiple path fading because of reflections and dispersions around and on the human body is accomplished in implanted interaction links that can adequately minimize the vitality and stability of interaction. To improve the accuracy of interaction in CWPCN systems, under the effect of every above-listed aspect, it is greatly suggested to employ a distinct method like MIMO.

This section explains the proposed workflow. This work proposed an implantable MIMO antenna in the IoT environment for wearable healthcare applications in CWPCN. Also, to enhance the capacity of the antenna a CSO technique is employed. Figure 7.2 portrays the flow of the proposed technique.

For the implementation, a four-component implantable MIMO antenna having EM band gaps is used, maximizing the rate of data transfer without employing extra spectrum or power transfer. MIMO channels are presented, providing substantial capacity gains than the conventional single-input single-output channels. Because of the distributed radiator framework, the areas of the suggested antenna are minimized. Decoupling is attained by lengthening a branch in the proportional axis of two propagation components, engraving a "+"-shaped channel, and inserting two EM band gaps in the radiator. Figure 7.3 displays the structure of the implantable MIMO antenna used for implementation. A compact MIMO antenna for wireless applications with band-notched characteristics is discussed [13].

The antenna is engraved on the commercially present Rogers 6010LM substrate having a thickness, h, width, W, and is enclosed by a superstrate with a similar substance used for the material. The four angles of the substrate are split into one-fourth rounded arcs, preventing pointed angles from

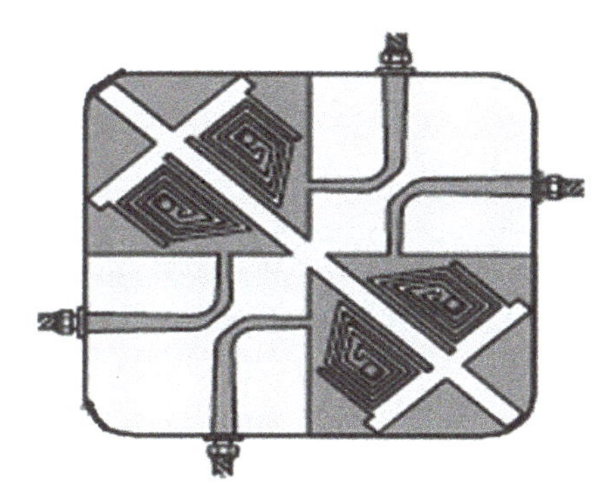

Figure 7.3 Structure of the four-unit implantable MIMO antenna.

damaging the tissues inside. Two radiators of square shape are engraved on the substrate, each one of which is consisted of double components present in the similar square radiator that is provided by double orthogonal diminished microstrips. The four components are equally reflected over double-slanted lines.

The vehicular communication using IoT system should be reachable, omnidirectional, and have negligible ground plane impacts. The auto business has shown expanded interest in different electronic mix issues, for example, over the ground radio wires, decreased link lengths, traveler well-being, and far-off keyless passage dependent on vehicular-to-everything correspondence, to further develop street security, traffic-less development, unwavering quality, and productivity. The unlicensed recurrence groups of 2.4 GHz and 5.470–5.725 GHz benefit wise transportation frameworks, while a recurrence band of 3.168–4.752 GHz offers multipath invulnerability for quick vehicles in thickly populated regions. Furthermore, the providing structures are perpendicular to one another, highly reducing the mutual coupling. A cross-shaped channel is engraved in the emitter and double EM band gaps are installed amid the radiator and the ground, maximizing the separation amid components.

7.4 RESULTS AND PERFORMANCE ANALYSIS

7.4.1 Low complex channel estimation model

Space-time block coding is a significant spatial and time diversity method that enhances the quality of the signal by employing uncomplicated action at the transmitter and successive decoding at the receiver. In this method, the representations S_0 and S_1 are transmitted concurrently from two transmit antennas T_{a1} and T_{a2} accordingly, at period t. At the successive period t +

T, T_{a1} sends $-S_1^*$ and T_{a2} sends S_0^*. The obtained signal is represented by the below equation:

$$R_s = HS + W \tag{7.1}$$

Here W denotes additive White Gaussian noise, and H that is represented by the common symbol $\{H_{ij}\}$ denotes the channels amid double transmit and receive antennas. The obtained signals at period t can be given by,

$$at\,Receiver1: Re_1 = S_0H_{11} + S_1H_{21} + W_1 \tag{7.2}$$

$$at\,Receiver2: Re_3 = S_0H_{12} + S_1H_{22} + W_3 \tag{7.3}$$

The signals obtained at time period $t + T$ can be given by,

$$at\,Receiver1: Re_2 = S_1^*H_{11} + S_0^*H_{21} + W_2 \tag{7.4}$$

$$at\,Receiver2: Re_4 = S_1^*H_{12} + S_0^*H_{22} + W_4 \tag{7.5}$$

W_1, W_2, W_3, and W_4 denote complicated Gaussian random variables denoting noise and intervention. The transferred representations S_0 and S_1 are calculated in a maximum probability pattern by initially merging the obtained signals corresponding to the subsequent equations

$$\widetilde{S_0} = H_{11}^*Re_1 + H_{21}Re_2^* + H_{12}^*Re_3 + H_{22}Re_4^* \tag{7.6}$$

$$\widetilde{S_1} = H_{21}^*Re_1 - H_{11}Re_2^* + H_{22}^*Re_3 - H_{12}Re_4^* \tag{7.7}$$

and then by employing a principle maximum probability identifier to try to retrieve S_0 and S_1 from $\widetilde{S_0}$ and $\widetilde{S_1}$.

The receiver needs information of the channel to retrieve the transferred signal effectively. In a training-dependent estimation of channel, in which a is the transferred training signal, the obtained signals are represented by

$$b = Ha + w \tag{7.8}$$

Here, w denotes the noise response. The channel response H is considered to be arbitrary and quasi-static inside the two blocks of transmission. Equation (7.8) is solved by reducing the cost function as

$$J(H) = (b - Ha)^H (b - Ha) \tag{7.9}$$

The gradient of Equation (7.9) is represented by the below equation:

$$\frac{\partial J(H)}{\partial H} = -2a^H b + 2a^H a \tag{7.10}$$

Figure 7.4 shows the MIMO wireless system model where the transmission and reception of data are processed.

Reducing the gradient to zero provides the calculation $\hat{H}$ of the channel response attained by

$$\hat{H} = \left(a^H a\right)^{-1} a^H b \tag{7.11}$$

The inversion of $a^H a$ has a maximum complication and shall substantially maximize when the count of transmit antennas maximizes. To prevent complications due to inversion of the matrix, orthogonal matrix triangularization is employed on a. In this technique, a sequence of reflection matrix is employed to the matrix, "a" column by column to eradicate the minimum triangular components. The reflection transformations are orthonormal matrices, which is represented by

$$A = \left(I + \gamma v v^H\right) \tag{7.12}$$

Here v denotes the Householder vector and $\gamma = -2v_2^2$. Figure 7.5 portraits the A_n construction steps.

The A_n constructed from the above steps are pre-multiplied by a consecutively as shown below:

$$A_k \ldots A_1 H = \begin{bmatrix} Q \\ 0 \end{bmatrix} \tag{7.13}$$

Here, Q denotes an upper triangular matrix, 0 denotes a null matrix, and the series of reflection matrices develop the complicated transpose of the orthogonal matrix P^H, that is, $P^H = A_k \ldots A_1$. Hence, Equation (7.13) can be represented by

$$a = P \begin{bmatrix} Q \\ 0 \end{bmatrix} \tag{7.14}$$

The error function for estimation of Equation (7.11) can be represented as

$$\varepsilon = b - \hat{H}a, if\ \varepsilon = 0, then\, b = \hat{H}a \tag{7.15}$$

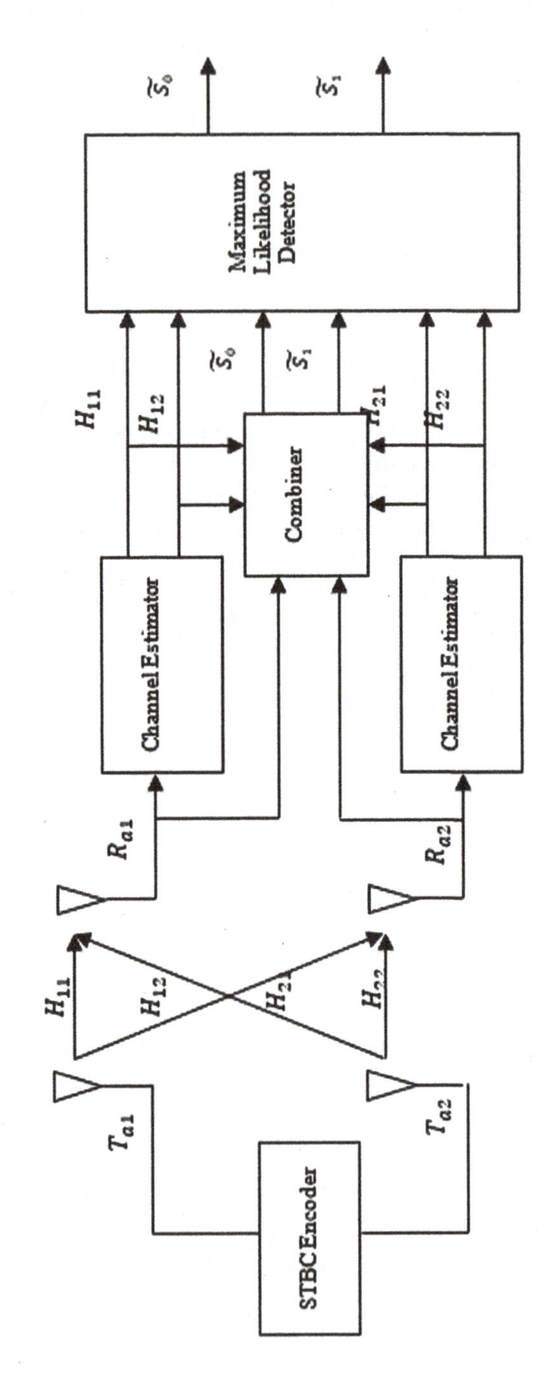

Figure 7.4 MIMO wireless system.

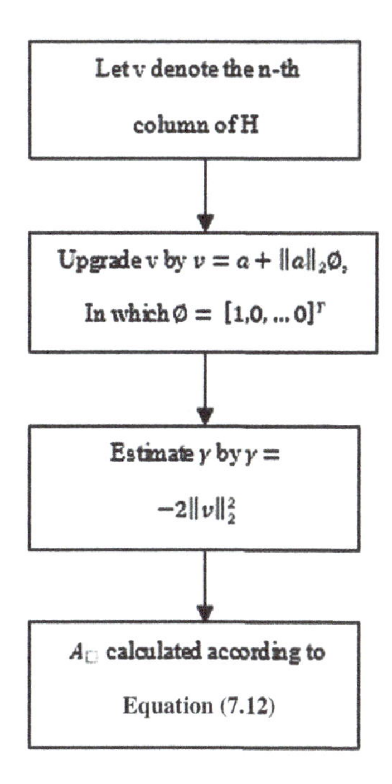

Figure 7.5 A_n construction steps.

By merging Equations (7.14) and (7.15), the obtained signal stands:

$$b = \hat{H}a = \hat{H}P\begin{bmatrix} Q \\ 0 \end{bmatrix} \tag{7.16}$$

The Hermitian of $P\begin{bmatrix} Q \\ 0 \end{bmatrix}$ is multiplied to both sides of Equation (7.16) to obtain the estimation of the channel:

$$\hat{H} = bP^H\begin{bmatrix} Q \\ 0 \end{bmatrix}^H \tag{7.17}$$

Since Q denotes an upper triangular matrix, $\hat{H}$ is resolved to employ back-substitution. This method confirms itself as an alluring solution of channel estimation, which leads to minimization of the complexity of the multiple antenna system. The sensors which play a vibrant role in actuating the systems take over the control of IoT and pass the data at every level through the fog system [14].

7.4.2 Data preprocessing

When several antennas are employed in one base station at a single cell, a few extra preprocessing of the basic channel information is required to create an essential input for any learning algorithm. An assumption is that a system in which channel state information acquisition occurs in the frequency area, compatible with the orthogonal frequency division multiplexing method and assumed as $H(c, t) \in \forall N$ represent the complicated direction denoting the spontaneous uplink distribution channel amid the single antenna user device and the N base station antennas on subcarrier(S) at period t. Although the location of the user device in the horizontal plane is fully fixed, the spontaneous channel as calculated by the baseband processor generally displays few changes because of limited fading and the remaining carrier resonance neutralization developing from the difficulty of absolutely integrating the oscillators of the transmitter and the receiver. To minimize the noise available in the incoming information and concurrently eliminate the general phase element because of the clock neutralization, we calculate the base station-side channel covariance as

$$C(x) = \mathbb{Z}[H(c,t), H(c,t)]^{\dagger} \tag{7.18}$$

for few static reference subcarrier $c$and in which $(.)^{\dagger}$ represents the Hermitian transposition. It is essentially considered that the prediction is calculated at period t above a minimum time horizon T when the user device stays static in a provided position x, hence the fading procedure is considered static.

By assuming a vectored and standardized variant of the covariance $C(x)$, as an input to the learning algorithm, represented by $y \in \mathcal{R}^{N^2}$. and obtained as

$$y = \left(vec_{ut}\left(Re\left\{ \frac{C(x)}{Tr(C(x))} \right\} \right)^T, vec_{ut}\left(Im\left\{ \frac{C(x)}{Tr(C(x))} \right\} \right)^T \right)^T \tag{7.19}$$

where $vec_{ut}(.)$ denotes the operator piling up in a direction the upper triangular items of a $N \times N$ matrix since they are nonzero and, thus, eliminates the basic repetition available in $C(x)$. . Such input selection arises from the insight that the second-order channel data acquire a majority of the position-associated features of the channel state information and, therefore, $C(x)$. is proportionally associated with the user equipment position x.

7.4.3 Chicken swarm optimization (CSO)

CSO is proposed for improving the capacity of the antenna by selecting the best channel for the transfer of the data [15]. CSO enhancement is an environment-stimulated meta-heuristics algorithm for enhancement, stimulated by the performance of chicken swarms consisting of roosters, hens, and chicks. Chicken swarms are divided into several groups depending on performance with every group containing rooster RO, hens HE, and chicks CH. By dividing swarm individuals, the performance analysis depends on fitness value. Chickens having worse fitness value are represented as CH, chickens having better fitness value are represented as RO and termed as dominant rooster whereas the rest of them are named as HE. Hens choose a group on their own for a living. Mother–child relation between hen and chick is estimated and mother hens are denoted as MH. Mother–child and dominance relation is static and the condition of each is upgraded for several time steps ts. Chickens track the rooster of the group when searching for food, whereas chicks track mother hen when searching for food.

7.4.4 Algorithm for CSO

Input the population of chicken with size N and set algorithm specifications such as G, size of RN, HN, CN, and MN;
Estimate the fitness values of every chicken, t 0, set up hierarchic order in the swarm and mother–child relation;
When (t<gen)
t =t + 1;
If (t %G==0)
Set up the hierarchic order in the swarm and mother–child relation;
Else
For i =1: N
If i==rooster
Upgrade the answer;
End if
If i==hen
Upgrade the answer;
End if
If i==chick
Upgrade the answer;
End if
Estimate the new answers;
Upgrade the new answers if they are excellent than the earlier ones;
End for
End if else
End while

7.4.5 Cooperative antenna selection – Neural network classifier

Convolutional neural network provides better outcomes in recent years in several areas associated with healthcare applications. The major advantageous factor of this type of classifier is minimizing the count of specifications in artificial neural networks [16]. This accomplishment has influenced both investigators and researchers to approach greater designs to resolve complicated works that were impossible with classic artificial neural networks.

The input of this classifier is a channel matrix H of the antenna system. The initial layer of convolution filters the input channel matrix with 32 kernels. Next, the initial pooling layer is presented to take the output of the initial convolution layer as input. It standardizes and pools the input into an output reaction. The max-pooling kernels possess a size of 2×2 and a stride of 2. The next convolution layer filters the input reaction with 64 kernels. Then, the next pooling layer transforms the input reaction into an output reaction. The max-pooling kernels possess a size of 2×2 and a stride of 2. The major concept of pooling is down-sampling to minimize the complication for upcoming layers. Pooling does not influence the count of filters. Max-pooling is the most general kind of pooling technique. Further, pooling is also utilized with non-similar filters to enhance efficacy.

The final layer is the fully connected layer which is a compact layer increasing the concurrence. It possesses 1,024 fully connected kernels of size 1×1. The ReLU layer is utilized as the activation operation of every convolution layer. The soft-max is used as the activation operation of fully connected layers. The cross-entropy was utilized as the loss operation.

To validate the antenna behavior within the human body, the implemented antenna is put in the multilayer tissue structure which comprises skin, fat, and muscle layers as portrayed in Figure 7.5. To estimate more approximately the actual body of the human and secure estimating time, a phantom of three layers is employed in the process of simulation. Figure 7.6 displays the scheme of embedding the implemented antenna in a three-layer phantom of the tissue that is comprised of skin, fat, and muscle to duplicate the original condition within the body of the human.

Table 7.1 shows the thickness of the human tissues and the electrical characteristics of the layers of the tissue that contains the numerical and the empirical phantoms.

A superstrate is employed to separate the radiator from the tissue, preventing being diminished and eroded, and additionally reduces the area of the implemented antenna. The implemented antenna is provided by four segments of 50-Ω coaxial cables. For simulation, the implemented antenna is embedded in the three-layer phantom as portrayed in Figure 7.6. The proposed methodology is contrasted with the existing methodologies

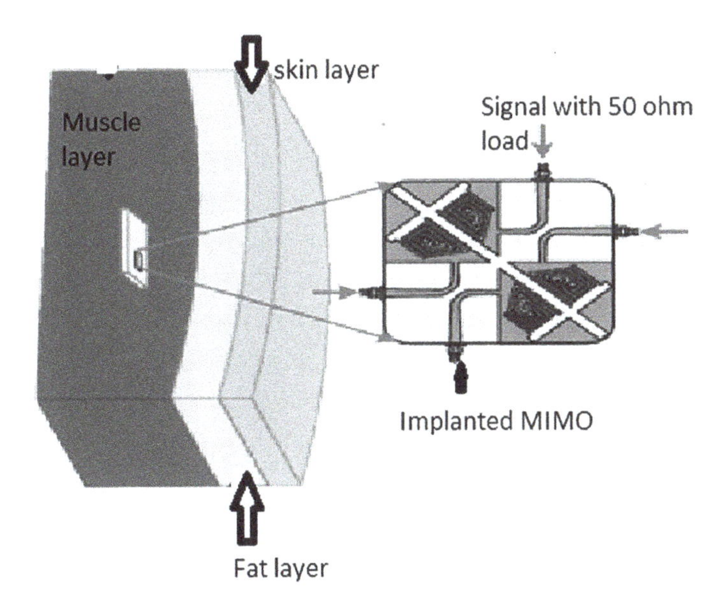

Figure 7.6 Three-layer phantom.

Table 7.1 Electrical characteristics of tissues at 2.45 GHz [17]

	Dielectric constant, ε_r		*Electrical resistivity σ (S/m)*	
Human tissues (thickness, mm)	*Numerical phantom*	*Empirical phantom*	*Numerical phantom*	*Empirical phantom*
Skin (1.5)	38.0	30.23	1.46	1.05
Fat (10.5)	5.28	4.85	0.104	0.091
Muscle (39.2)	52.73	51.29	1.74	2.07

concerning gain, mutual coupling, specific absorption rate (SAR), bandwidth, and efficiency.

7.4.6 Gain

Gain explains the amount of power that is transferred along with the orientation of maximum radiation to that of an isotropic source. Figure 7.7 shows the comparison of gain in dBi for the existing and the proposed method [18]. The suggested technique surpasses the traditional techniques concerning overall gain.

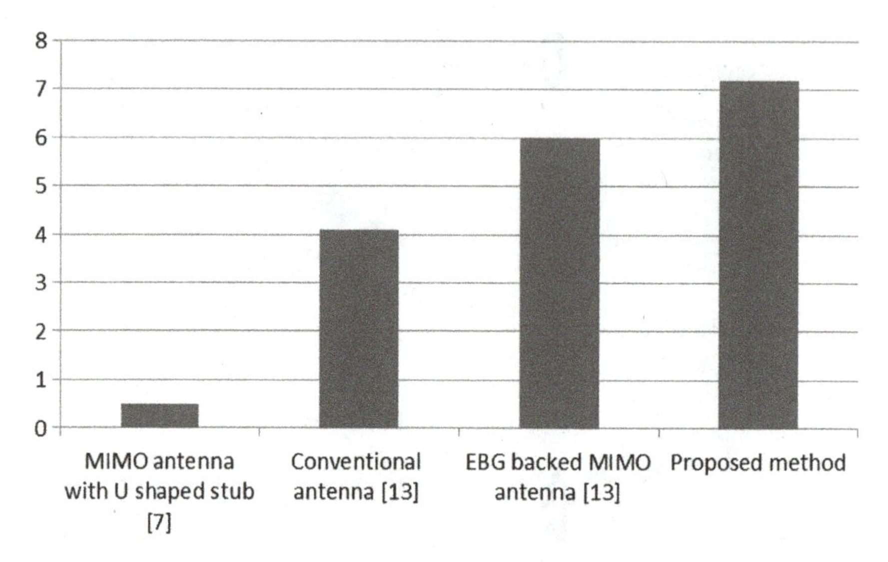

Figure 7.7 Comparison of gain (dBi) for existing vs. proposed method.

7.4.7 Mutual coupling

Mutual coupling amid the antenna components affects the performance of the antennas. Figure 7.8 portrays the comparison of mutual coupling in –dB for the existing and the proposed methods. The mutual coupling for the proposed method is less when compared to that of the existing methods, which improves the performance of our method.

7.4.8 Specific absorption rate (SAR)

The radiations in the near field of the antennas that are wearable influence the body of the human. Simultaneous disclosure to the radiations maximizes the heat level of the tissues in the human body. Heat is produced because of the non-radiating sensitive near fields. Excess heat shall affect the tissues and minimizes the flow of blood leading to a disruption in the operation of delicate organs. Hence, it is essential to take into account the power consumed by tissues of the human body. A standard rule to estimate the consumed EM power by the body is termed the SAR. As stated in the IEEE C95.1-2005 benchmark, the number of SAR must not exceed 2 W/kg balanced over 10 g of tissue. Simulated SAR for the implanted MIMO antenna is displayed in Figure 7.9. The antenna possesses a higher SAR value of 0.457 W/kg for 0.1 W of input power. The attained number is sufficiently less than the highest security restraint.

Figure 7.10 portrays the comparison of SAR for the existing and suggested method for 0.1 W of power input. The SAR value of the proposed method is

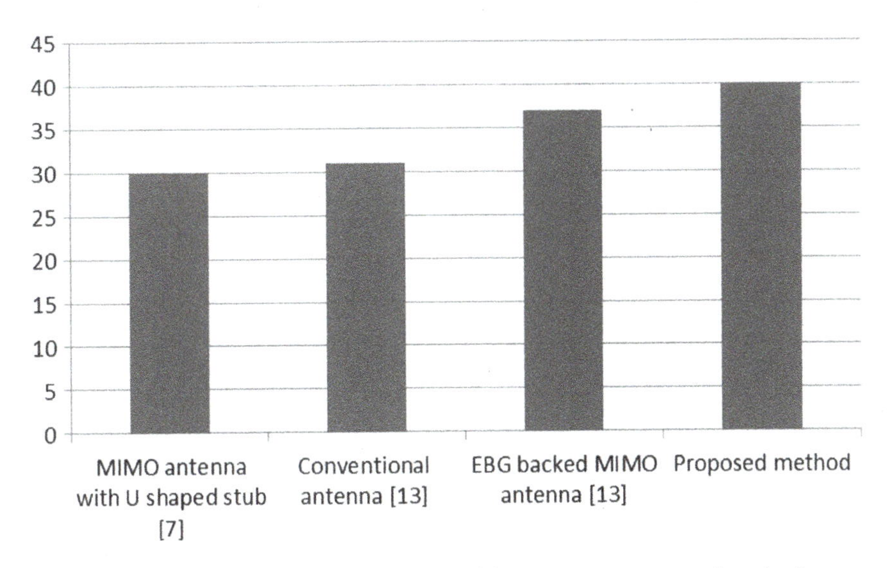

Figure 7.8 Comparison of mutual coupling (–dB) for existing vs. proposed method.

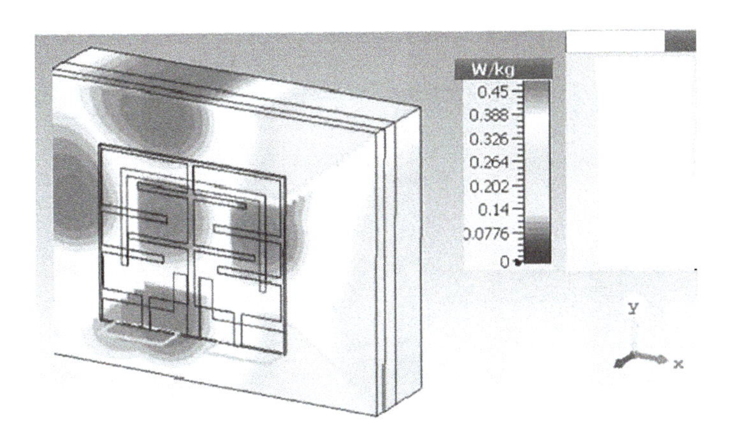

Figure 7.9 SAR of the implemented MIMO antenna.

lower than the existing method which minimizes the damage to the tissues of the human body.

7.4.9 Bandwidth

The bandwidth is the transmission capacity of an antenna and is an important aspect when estimating the quality of an antenna. It is used to describe the

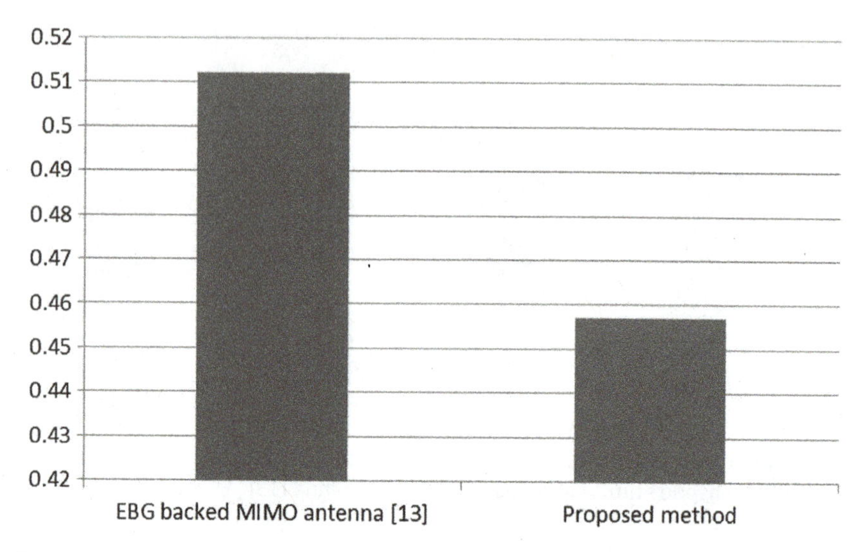

Figure 7.10 Comparison of SAR (W/kg) for existing vs. proposed method.

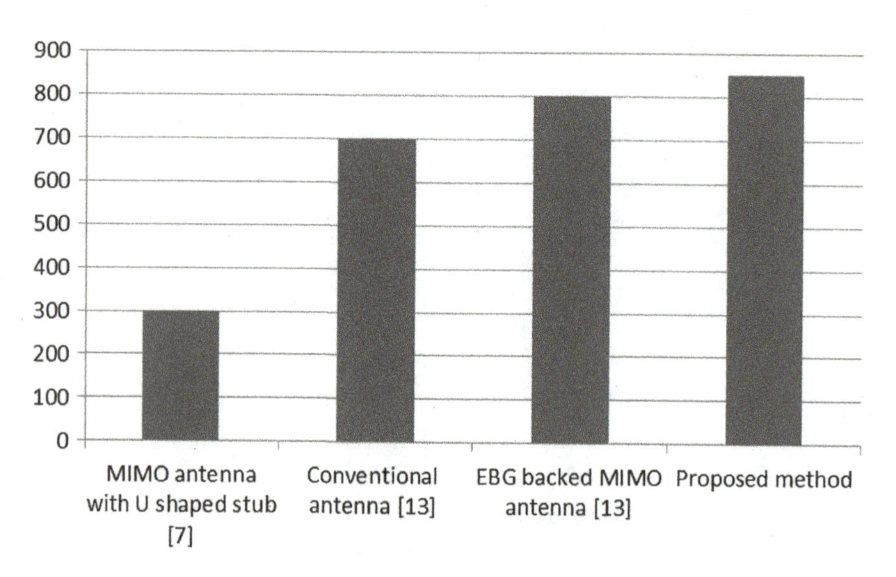

Figure 7.11 Comparison of bandwidth (MHz) for existing vs. proposed method.

variation between the upper and lower frequencies in transmission such as a radio signal and is typically calculated in hertz (Hz). Figure 7.11 shows the comparison of bandwidth of existing and proposed methods. It is observed that the suggested method has an improved bandwidth when compared with the traditional ones.

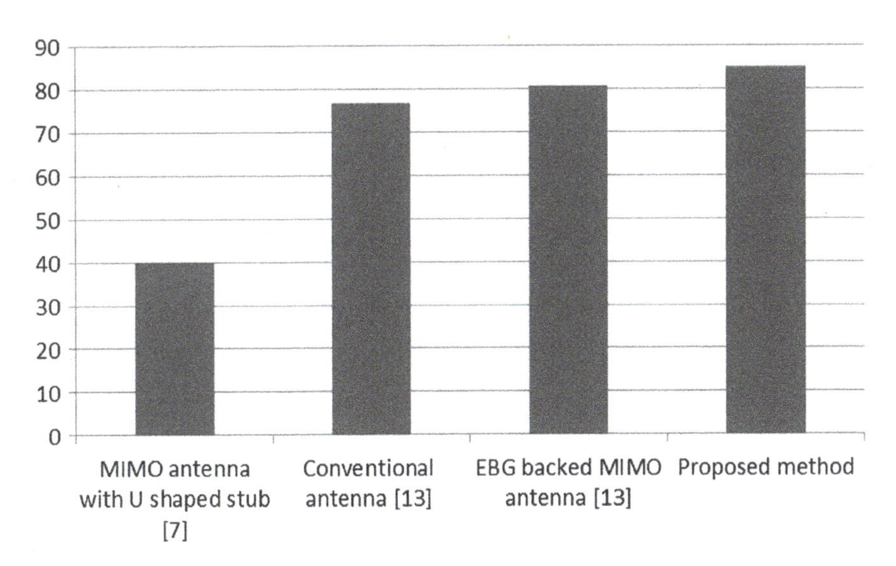

Figure 7.12 Comparison of efficiency (%) for the existing and proposed method.

7.4.10 Efficiency

Figure 7.12 shows the comparison of the efficiency of the existing and the proposed method. It is clear from the graph that the suggested method is better for wearable healthcare applications when compared with the existing methods.

7.5 CONCLUSION AND FUTURE DISCUSSION

A small four-unit MIMO antenna functioning in the 2.45-GHz ISM band is implemented in the IoT environment for wearable healthcare applications in CWPCN. The antenna capacity is improved by the chicken swarm secrecy probability optimization method and classification was done using the cooperative antenna selection – neural network classifier. The antenna is examined on various parts of the body of the human using a three-layer phantom and the results are investigated. The behavior of the proposed method is analyzed concerning bandwidth, efficiency, gain, mutual coupling, and SAR and compared with the existing methods. It is concluded that the suggested method is the best for flexible and wearable healthcare devices in terms of bandwidth, efficiency, gain, mutual coupling, and SAR values within mentioned input power. The SAR result of the implanted MIMO antenna ensures that the antenna behaves satisfactorily and is relevant for wearable applications, and it works within the agreeable limits of SAR as stated by Federal Communications Commission (FCC). The proposed method is competing for wearable utilizations because of the antenna's compact size, single-layer framework, simple incorporation, reliability, and reasonable on-body antenna gain.

REFERENCES

[1] Bocan, K.N., Mickle, M.H. and Sejdić, E., 2017. Tissue variability and antennas for power transfer to wireless implantable medical devices. *IEEE Journal of Translational Engineering in Health and Medicine*, 5, pp.1–11.

[2] Kulaç, S., Sazli, M.H. and Ilk, H.G., 2018, September. External relaying based security solutions for wireless implantable medical devices: A review. In 2018 11th IFIP Wireless and Mobile Networking Conference (WMNC) (pp.1–4). IEEE.

[3] Muthu, B., Sivaparthipan, C.B., Manogaran, G., Sundarasekar, R., Kadry, S., Shanthini, A. and Dasel, A., 2020. IoT-based wearable sensor for disease prediction and symptom analysis in the healthcare sector. *Peer-to-Peer Networking and Applications*, 13(6), pp.2123–2134.

[4] Goyal, S., Sharma, N., Bhushan, B., Shankar, A. and Sagayam, M., 2021. IoT enabled technology in secured healthcare: Applications, challenges, and future directions. In Cognitive Internet of Medical Things for Smart Healthcare (pp.25–48). Springer, Cham.

[5] Zhou, Z., Yu, H. and Shi, H., 2020. Human activity recognition based on improved Bayesian convolution network to analyze health care data using wearable IoT device. *IEEE Access*, 8, pp.86411–86418.

[6] Faisal, F., Zada, M., Ejaz, A., Amin, Y., Ullah, S. and Yoo, H., 2019. A miniaturized dual-band implantable antenna system for medical applications. *IEEE Transactions on Antennas and Propagation*, 68(2), pp.1161–1165.

[7] Gupta, A., Kansal, A. and Chawla, P., 2021. Design of a wearable MIMO antenna deployed with an inverted U-shaped ground stub for diversity performance enhancement. *International Journal of Microwave and Wireless Technologies*, 13(1), pp.76–86.

[8] Supriya, A., Kumar, S.A. and Shanmuganantham, T., 2020, July. Design of CPW fed antenna with split ring resonator for ISM band for biomedical applications. In 2020 IEEE International Conference on Electronics, Computing and Communication Technologies (CONECCT) (pp.1–3). IEEE.

[9] Bhuvaneswari. M. and Sasipriya. S 2019. Performance analysis of various techniques adopted in a 5G communication network with a massive MIMO system, Proceedings of the 5th International Conference on Inventive Computation Technologies, ICICT 2020, 2020, pp.283–286, 9112446.

[10] Sun, G., Zhao, X., Liang, S., Liu, Y., Zhou, X. and Zhang, Y., 2019, September. A modified chicken swarm optimization algorithm for synthesizing linear, circular, and random antenna arrays. In 2019 IEEE 90th Vehicular Technology Conference (VTC2019-Fall) (pp.1–7). IEEE.

[11] Arun Chakravarthy, R. Arun, M. Sureshkumar, C. and Bhuvaneswari, M., 2021. Waste management solution for smart city using Internet of Things. *International Journal of Creative Research Thoughts*, 9(2), pp. 3908–3913.
[12] Arun Chakravarthy, R. Bhuvaneswari, M. and Arun, M., 2020. IoT based environmental weather monitoring and farm information tracking system, *Journal of Critical Reviews*. 7(7), 307–310.
[13] Kumar, S., Kumar, R., Vishwakarma, R.K. and Srivastava, K., 2018. An improved compact MIMO antenna for wireless applications with band-notched characteristics. *AEU-International Journal of Electronics and Communications*, 90, pp.20–29.
[14] Bhuvaneswari, M. Arun Chakravarthy, R. and Arun, M., 2020. Low cost design automated adhesive dispenser for industry, *Journal of Critical Reviews*. 1(12), 248–251.
[15] Deb, S., Gao, X.Z., Tammi, K., Kalita, K. and Mahanta, P., 2020. Recent studies on chicken swarm optimization algorithm: A review (2014–2018). *Artificial Intelligence Review*, 53(3), pp.1737–1765.
[16] Cai, J.X., Zhong, R. and Li, Y., 2019. Antenna selection for multiple-input multiple-output systems based on deep convolutional neural networks. *PloS one*, 14(5), p.e0215672.
[17] Fan, Y., Huang, J., Chang, T. and Liu, X., 2018. A miniaturized four-element MIMO antenna with EBG for implantable medical devices. *IEEE Journal of Electromagnetics, RF and Microwaves in Medicine and Biology*, 2(4), pp.226–233.
[18] Iqbal, A., Basir, A., Smida, A., Mallat, N.K., Elfergani, I., Rodriguez, J. and Kim, S., 2019. Electromagnetic bandgap-backed millimeter-wave MIMO antenna for wearable applications. *IEEE Access*, 7, pp.111135–111144.

Chapter 8

Intelligent traffic light systems for the smart cities

Rajiv Dey

CONTENTS

8.1 INTRODUCTION

According to the prediction carried out by Urban Organization, a proprietary of the United Nations around 68% of the world's total population will live in the urban areas by 2050. This unplanned city growth lead to a multiple number of issues to the population such as enhanced traffic congestion, environmental pollution, accidents, and so on. To solve such issues related to the urban growth, numerous number of electronic system appeared across the world, each focuses on solving a particular issue of everyday life. Moreover, in the past few decades substantial industry and academic efforts have been put on facilitating the traffic pressures and on mitigating traffic flow pressures. Over the past few years, the traffic signal control has advanced a lot from the Information and Communication Technology. The Internet of Things (IoT) is one such technology by virtue of which solutions have been developed to solve real-life challenges [1]. The places where such technologies are implemented to make that place smart are known as smart cities. Figure 8.1 represents a basic composition of a smart city. In

DOI: 10.1201/9781003230236-8

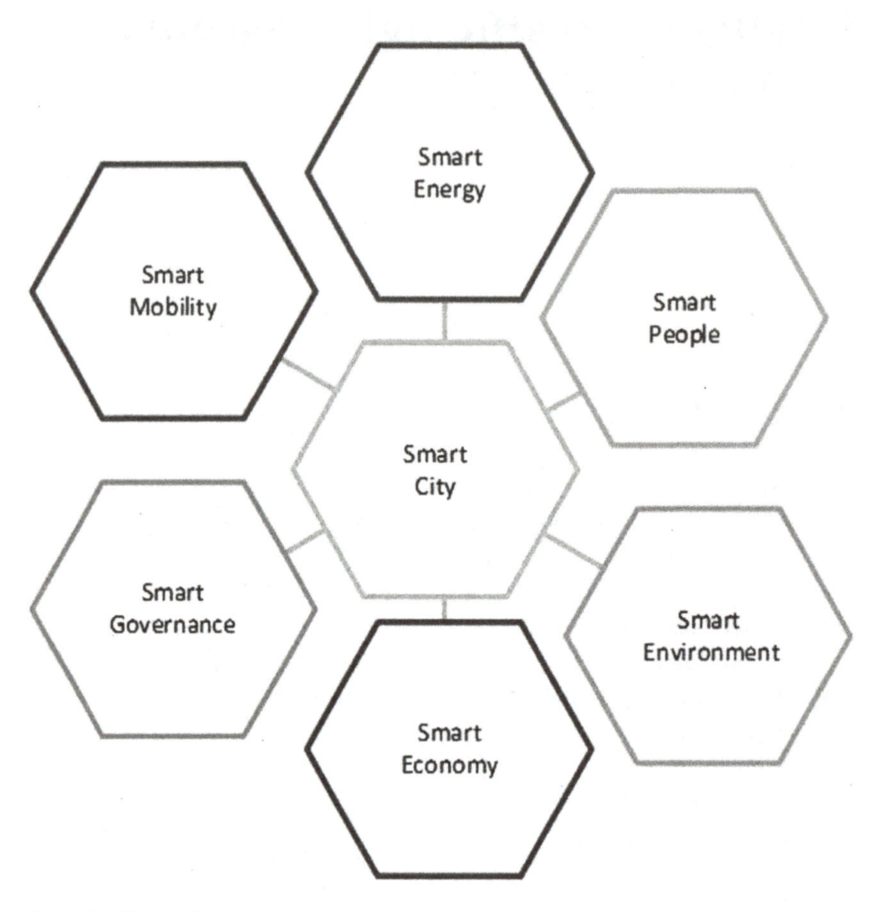

Figure 8.1 Smart city composition.

Figure 8.1, it can be seen that various composition sectors of smart suggest specific solution to a real-life problem of urban areas.

Smart mobility is one of the potential application areas where traffic congestion nowadays is a major issue which can lead to loss of time, environmental pollution, and accidents [2]. In such congested networks, the vehicle movements are handled by traffic lights and are moreover exaggerated by route selections that the drivers make. Therefore, the traffic situations can thus become better by the advancement of disciplined traffic signal control and route selection methods [3]. Hence, the technical solutions in this area can minimize traffic congestion, accidents, and environmental and noise pollution. It is mentioned in Schrank et al. 2014 that the North Americans had an additional overhead of 6.9 billion hours of travel, due to which an extra 3.1 billion liters of fuel is consumed.

This results in a bottling cost of around 160 billion dollars. In a conventional system, an optimum predetermined schedule has been followed by the traffic light controllers, which in turn result in a deprived performance under time-varying traffic environments and very high traffic demands. This issue can be eased through the integration of traffic signal with adaptive controllers, such as SCATS, SCOOT, UTOPIA, and so on. This chapter is written with the endeavor to deliver the readers a review on the recent advancement in the domain of intelligent traffic light system (ITLS) to solve real-life issues.

8.2 TRAFFIC LIGHT CONTROL SYSTEM

The word traffic light originates from the Greek word "Semaphoros" which is a combination of sema and phoros: sema means light and phoros means a signaling apparatus [4]. John Peake Knight is the inventor of the world's first traffic light signal controller installed at crossroad city London in 1868 [5]. From the past to the present, the purpose of the traffic light controller remains same, that is, to control the flow of traffic. Whereas the way by which traffic flow is managed has been changed significantly [6].

8.2.1 Conventional traffic light system

The conventional traffic light system runs on the prefixed time base that triggers the signal lights manually. The configuration of the conventional traffic light system can be carried out by an expert in that domain or it can be done using wire connected network [7] which interconnects all the traffic lights of the town to a solo point. Following are the advantages of the traditional traffic light system:

i. Due to the use of only a time-based circuit to drive the lights, this system is simple in nature.
ii. Wired connection between the traffic lights and the centralized system ensures the dedicated connection between them to communicate with each other.

Despite being simple, the conventional traffic light system has the following disadvantages:

i. It does not consider the present traffic flow on the roadside and therefore, the switching timing of traffic lights is not optimized which result in wastage of time, sudden traffic jams, and accidents in few cases.
ii. Despite of ensuring cable connections between the traffic signals within a network, the cable length in such networks can range in kilometers for such urban area networks, and the most of the traffic

signal malfunctions occur only due to wiring damage and adverse effect of temperature and humidity on the cables [4].

8.2.2 Intelligent traffic light system

To overcome the issues encountered by the conventional traffic light system, the concept of an ITLS came into picture. The ITLS is the electronics-driven device that manages the pedestrian and traffic dynamically by measuring the real-time parameters such as vehicle count on a particular road and the speed of vehicles. These devices acquire the real-time data using sensors such as infrared sensor [8], magnetic sensor [9], cameras, and so on. By knowing the vehicle count passing through the crossing and the speed of the vehicle, efficient time schedule algorithms can be implemented and the time needed by each vehicle to reach the next crossing can also be calculated [8, 10]. Figure 8.2 shows the ITLS with various advancements.

The intelligent traffic lights have the ability to report the status of traffic lights, that is, any faults in the traffic lamps, communication or control failure allows for the immediate attention of control station. Moreover, eliminating the issues related to the wired connections present in the conventional traffic lights, the intelligent traffic lights utilizes wireless communication technologies such as ZigBee, radio frequency (RF), Wi-Fi, and so on.

ITLS is a subset of intelligent transportation system, and it is a complex system that integrates numerous IT technologies which includes traditional IT technologies such as industry and IT technology and the emerging IT technologies such as intelligence-based technology. Among all these technologies, IoT plays a crucial role for collecting big data, analyzing big data, and managing big data. By linking ubiquitous devices and services by numerous networks, IoT demonstrates an encouraging future to deliver efficient and safe facilities for all applications, nearly anytime and anywhere. In nearly every situations, IoT is considered as a new cohort of Internet; it is due to the fact that it enlarges the communication from humanoid and humanoid to humanoid and things or things and things. With the inclusion of sensors in such IoT-enabled ITLS, physical transportation systems can be effectively observed in real time, and a massive volume of data is produced. The data obtained from traffic light system are acquired from physical sensors, such as global positioning system, induction coil, video camera, and so on have been extensively utilized in transport management, which conveyed great comfort to our lives. However, the sensors that were used in these systems also suffer from some shortcomings, particularly for specific applications. For example, the data obtained from a floating car is inadequate, as the cab in one metropolitan is limited and cannot cover each road segment concurrently. Moreover, the life of each induction coil is typically not more than two years and there is still much scope for the enhancement of the reliability

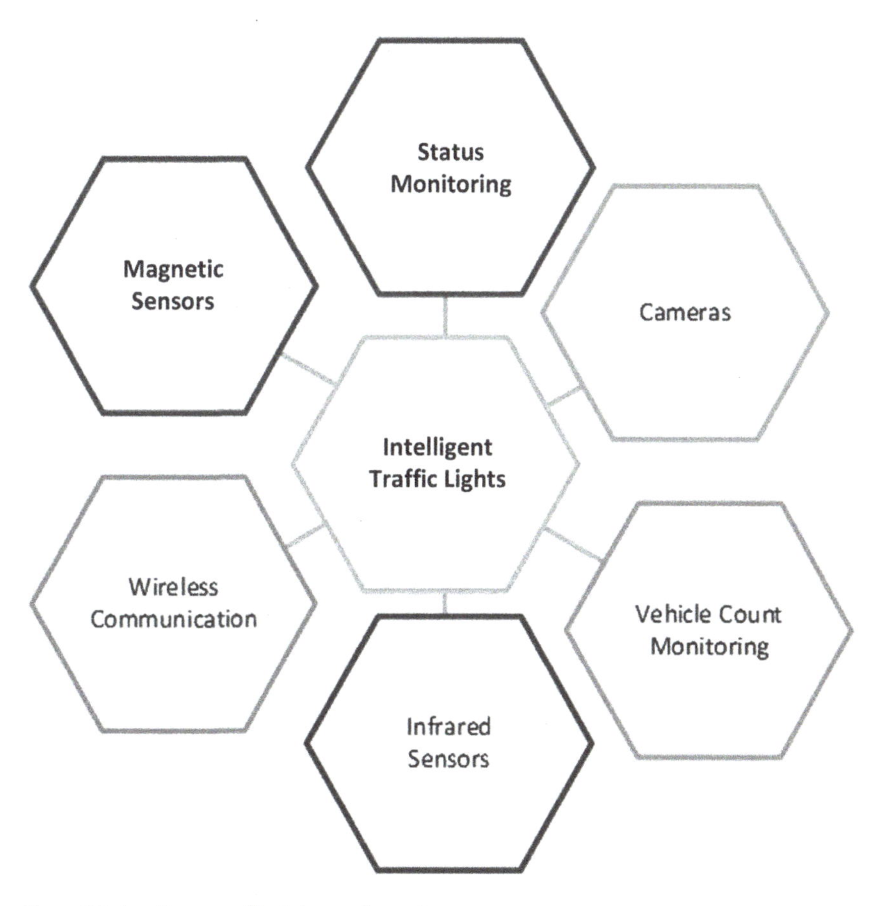

Figure 8.2 Intelligent traffic light configuration.

of such kind of sensors; it is also hard to get clear images in argumentative weather and stumpy light conditions.

8.3 BASIC ELECTRONIC CIRCUIT OF INTELLIGENT TRAFFIC LIGHT SYSTEM

An elementary ITLS comprises a microcontroller-based electronic circuit (able to switch AC/DC lamps, check the status of lamps) and a wireless communication module for transmitting data to a remotely situated control station. The basic building blocks of an ITLS are shown in Figure 8.3. The entire system is controlled and operated using a central Advanced Risc Machine (ARM) microcontroller where all the coding parts for implementing the efficient algorithms have been done. To operate all the crossing in real

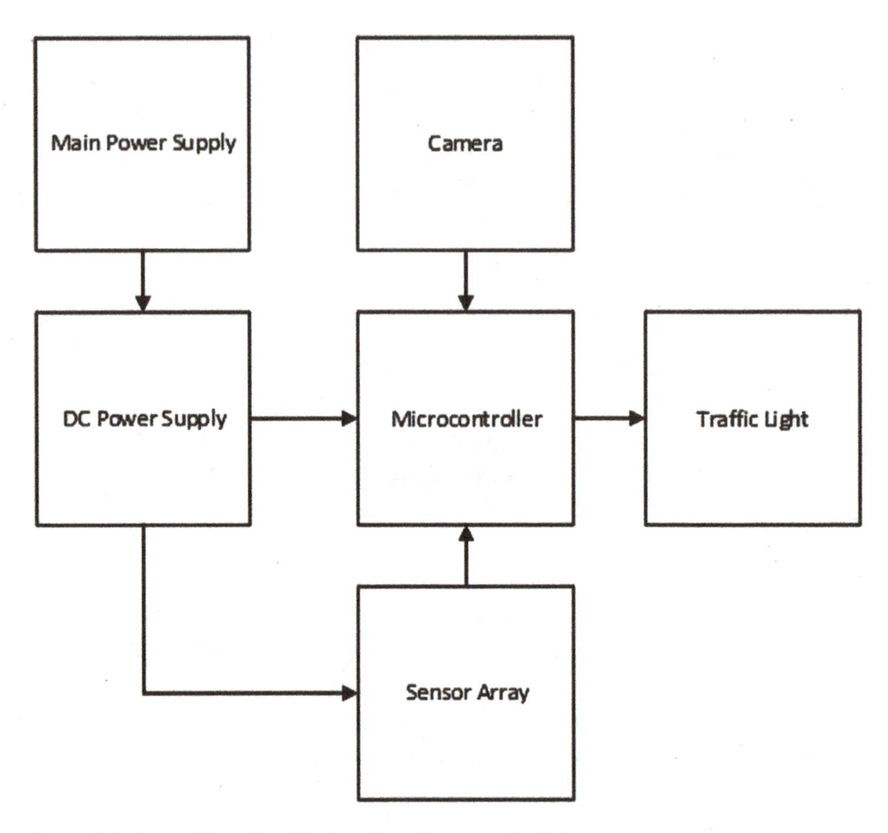

Figure 8.3 General building blocks of intelligent traffic light system.

time and in synchronization, a real-time clock is used as a reference clock source for all the traffic light systems in a network.

8.4 WIRELESS COMMUNICATION NETWORK-BASED CENTRALIZED TRAFFIC MANAGEMENT SYSTEM

Several communication technologies arise for doing wireless communication between the traffic light systems depending upon the operational circumstances. However, for the cases where a large number of devices are present in a network with different data rates and range requirement, then the choice to choose a particular technology among many is restricted. Mesh network is considered to be one of the ideal networking topologies in smart cities by the researcher community. This is due to the fact that the mesh network has the capability to regenerate itself in the face of any interruption from the network. Several works reported in the literature in

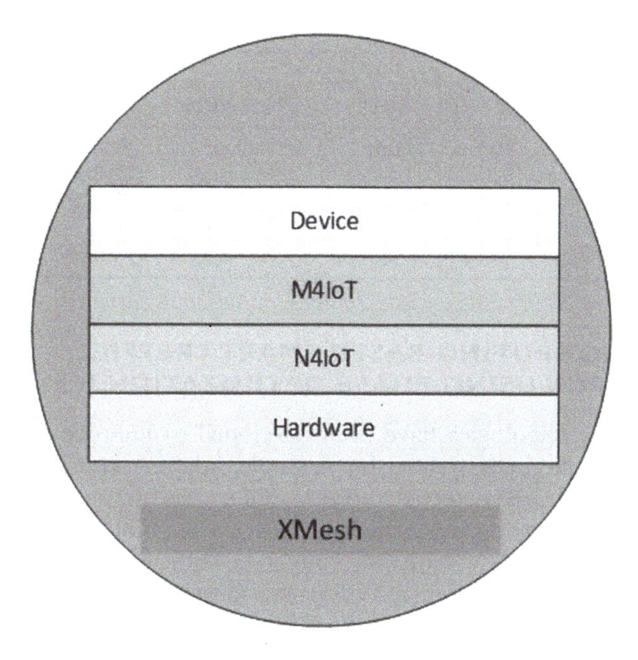

Figure 8.4 XMesh network model.

which the mesh networks are created using ZigBee protocol [11, 12]. The mesh network created using ZigBee is popularly known as Xmesh originally developed as a part of doctoral research work in the University of Campinas (commonly known as Unicamp). Figure 8.4 shows a layered architecture of an Xmesh network.

In Figure 8.4, two blocks are added in between the device and hardware layers such as N4IoT (network layer having mathematical modelling for routing) and M4IoT (a protocol used to create virtual private device network). It utilizes an encryption algorithm named elliptic curve for security purposes with less payload compared to the other mesh network technologies.

The Xmesh network works at 2.4 GHz ISM band (industrial medical and scientific band) and supports a data rate of 2 Mbps. It provides two options for the output power depending upon the range of coverage. For 100 m distance, the output power utilized is 0 dBm and for 1,800 m it is 20 dBm. At shorter distances, the traffic signals only exchange messages with the other traffic lights. Each data frame comprises 30 bytes, 20 bytes for the header, and 10 bytes for the data. A comparison between the Xmesh network and other wireless networking technologies is shown in Table 8.1.

Table 8.1 Wireless networking technologies comparison

Technology	*Supported data rate*	*Range of coverage*	*Power*	*Network size*	*Technology*	*Supported data rate*
WiFi	1.3 Gbps	100 m	High	Medium	Wi-Fi	1.3 Gbps
ZigBee	250 kbps	60 m	Low	Very large	ZigBee	250 kbps
Bluetooth	2.1 Mbps	100 m	Low	Small	Bluetooth	2.1 Mbps
LoRa	100 kbps	5 km	Very low	Medium large	LoRa	100 kbps
XMesh	2 Mbps	1,800 m	Medium	Very large	XMesh	2 Mbps

8.5 FOG COMPUTING-BASED SMART TRAFFIC CONTROL USING PHASE OPTIMIZATION [13]

Several latest technologies have been developed to improve the performance of the traffic light system, for example, adaptive traffic light system which utilizes infrared detector, cameras, radars, and so on to acquire real-time traffic data to resolve challenges such as unpredictable traffic flow, blending the traffic data with IoT (might be difficult due to the heterogeneity embedded in the vehicles). Recently developed technology such as Vehicular Ad-Hoc Network and the IoT makes the data acquisition and monitoring in adaptive traffic light system possible [13]. Greater connection capability along with the sensing and data acquisition enhances the amount of data to be processed. Therefore, it becomes necessary to optimize the phase timing depending on the acquired data. Moreover, the present control approaches need to improve the long decision and latency issues. It is observed that the traffic flow has already been changed before the result of optimized traffic data reached the traffic control system. Therefore, it can be concluded that the optimized traffic control strategy does not fit for the current traffic conditions. To overwhelmed this issue, a novel fog computing-based strategy may be a possible solution [14]. The fog computing is one of the recently introduced technologies for intelligent transportation and smart cities [15, 16].

The fog computing brings the data storage and processing power closer to the data generation (edge device), enhancing the computational capability while maintaining the latency constraints. The fog device processes all the acquired data related to different surface and traffic conditions and instantly responds to the traffic control system to avoid the traffic congestion and ensure driver safety. Works done in this area consider the roadside unit (in short RSU) as a distinct fog node to perform fog commutating. The other role of the fog node is to interact with the vehicles for data propagation and information exchange, and most of the data processing and execution of algorithms are done in the fog node itself to obtain an optimal schedule for traffic shaping. Some works consider assimilation of V2X technology to achieve optimum performance, for example, to achieve greater efficiency in traffic control plan, Priemer et al. [17], acquires the speed, acceleration, and vehicles heading data using V2I communication technology at the intersections. Furthermore, a vehicular fog computing architecture has been proposed in Hou et al. [18]; this work takes benefit of cooperative gathering of end user customers and the surrounding end node devices to perform computation and communication tasks. Particularly, individual vehicle either in stopped or moving condition acts like a fog node to acquire the desired traffic data. For some emergency cases such as taking a patient to a nearby hospital in an ambulance in a minimum span of time are a condition of survival and death. In such circumstances, fog and cloud computing can collaboratively play an important role in achieving quick arrival of the ambulance to the hospital. The fog computing for instance can obtain the momentary phase timing plan so that at the intersection, the ambulance can easily pass without waiting for the traffic signal to get cleared. After that, the traffic control can resume the previous timing schedule. The fog and cloud computing collectively enhance the running efficacy of ambulances to a great extent.

In general, the fog computing-based phase optimization consists of three layers, namely

i. Cyber physical layer
ii. Fog computing layer
iii. Cloud computing layer

i. Cyber physical layer: This layer consist of a densely distributed network of sensors such as loop sensors and surveillance cameras fitted on the roadsides. The vehicles having the communication devices on board or the roadside communication devices communicate with each other for forwarding and gathering the data. This layer is accountable for gathering the traffic-related data and forwarding it to the fog node for further processing. However, the traffic data heterogeneity from a variety of sensors creates novel issues related to data extraction, fog computation, and storage. The other matter of concern in communication between the sensors and actuators other than 3G, 4G, Bluetooth, Wi-Fi, and ZigBee is that all these communication technologies can provide both V2V and V2I communications.

ii. Fog computing layer: A block diagram representation of the fog computing layer is shown in Figure 8.5. The fog node consists of the computation and storage facility close to the traffic light control system (edge node). The fog node is basically responsible for supervising the data handling which includes storage, computation, and event management such as emergency service call and traffic signal optimization. The fog architecture comprises numerous modules; each module plays a different function in optimization

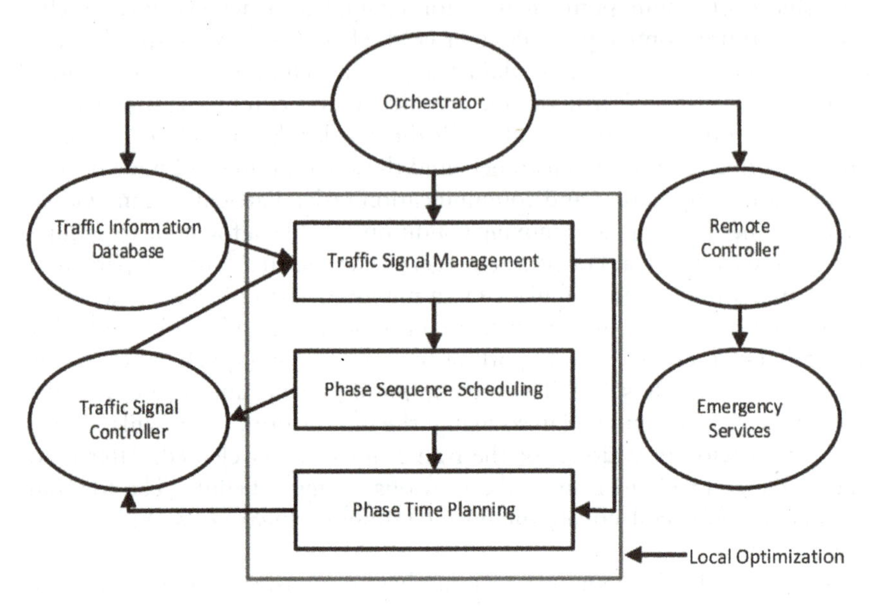

Figure 8.5 Layered architectural view of fog computing.

of the phase timing of the traffic signal plan. The functionality of each fog layer architecture is as follows:

a. Traffic information database: It is responsible for storing the traffic information data and further uploading it to the cloud.
b. The Orchestrator model: This model is accountable for assessing the traffic status and triggering event handler function. In the course of traffic congestion, it can instantiate the local optimization procedure.
c. Local optimization block: The rectangular block in Figure 8.5 represents the local optimization module. In this block, the traffic data is gathered from the traffic information database and depending upon the situation, it uses a specific algorithm to plan the phase sequence and timing for optimized performance. The resultant phase timing schedule is forwarded further to the traffic controller responsible for regulating the traffic light timings. In the case of emergency situations such as accidents of vehicle collision the Orchestrator model will need direct help in the form of human intervention in the traffic control center. The fog computation has numerous advantages such as reduced latency in data processing and response. Therefore, the traffic flow can be managed efficiently within the minimum amount of time.

iii. Cloud computing layer: Compared to the fog layer, this layer is associated with a large number of powerful computing and storage facilities. The main difference that separates the fog layer from the cloud layer is the following: the cloud layer serves multiple number of intersections simultaneously, whereas the fog layer delicately serves only single intersection. Built on the filtered and processed data of the fog layer, the cloud layer basically focuses on upholding regional synchronization among the intersection in a region. It is responsible for optimizing the phase timing of the traffic control system globally, whereas the fog layer optimizes only the phase timing plan of single intersection. Moreover, the fog computing is not accountable to coordinate between the intersections present before and after it with the upstream and downstream data. Therefore, it can be concluded that the traffic control at the fog nodes is not globally optimal. For more insights on this topic, the readers are encouraged to read the article [13].

8.6 CONCLUSIONS

Nowadays, traffic congestion has become a critical issue on the schedule of various public/private stakeholders due to the continually rising metropolitan traffic volumes and the lack of space, and public funds to build new transport foundations. These issues are combined with the complexity of knowledge, modeling, and managing the dynamics of traffic networks. ITLS has been appeared as a significant component and widely accepted for the smart city as it overcomes the shortcomings of the conventional traffic light system. This chapter tries to explore the recent advancements done by the researchers in the domain traffic light system to solve the real-life challenges of traffic congestion, fuel wastage, and environmental pollution.

REFERENCES

[1] N. Chandrashekharan, “Special Report: The Internet of Things,” 2014. http://theinstitute.ieee.org/%0Astatic/special-report-the-internet-of-things

[2] H. Frumkin. 2018, “Urban Sprawl and Public Health,” *Association of Schools of Public Health Digitaljournals.Moph.Go.Th,* vol. 117, no. 3, pp. 201–217, 2015, [Online]. Available: www.jstor.org/stable/4598743

[3] H. Kamal, M. Picone, and M. Amoretti, “A Survey and Taxonomy of Urban Traffic Management: Towards Vehicular Networks,” pp. 1–40, 2014, [Online]. Available: http://arxiv.org/abs/1409.4388

[4] Semaphore, “English Oxford Living Dictionaries,” 2013. Available: https://en.oxforddictionaries.com/definition/%0Asemaphore

[5] BBC, “The Man Who Gave Us Traffic Lights.” Available: www.bbc.co.uk/nottingham/content/%0Aarticles/2009/07/16/john_peake_knight_traffic_lights_%0Afeature.shtml

[6] L. F. P. de Oliveira, L. T. Manera, and P. D. G. D. Luz, “Development of a Smart Traffic Light Control System with Real-Time Monitoring,” in *IEEE Internet of Things Journal*, vol. 8, no. 5, pp. 3384–3393, 1 March 2021, doi:10.1109/JIOT.2020.3022392

[7] O. Younis and N. Moayeri, "Employing Cyber-Physical Systems: Dynamic Traffic Light Control at Road Intersections," *IEEE Internet Things Journal*, vol. 4, no. 6, pp. 2286–2296, 2017, doi:10.1109/JIOT.2017.2765243

[8] L. Paul Jasmine Rani, M. Khoushik Kumar, K. S. Naresh, and S. Vignesh, "Dynamic Traffic Management System Using Infrared (IR) and Internet of Things (IoT)," in *2017 Third International Conference on Science Technology Engineering & Management (ICONSTEM)*, 2017, pp. 353–357, doi:10.1109/ICONSTEM.2017.8261308

[9] Qing Wang, Jianying Zheng, Hao Xu, Bin Xu, and Rong Chen, "Roadside Magnetic Sensor System for Vehicle Detection in Urban Environments," *IEEE Transactions on the Intelligent Transportation Systems*, vol. 19, pp. 1365–1374.

[10] Wei Zhou, Lin Yang, Tianxing Ying, Jingni Yuan, and Yang, "Velocity Prediction of Intelligent and Connected Vehicles for a Traffic Light Distance on the Urban Road," *IEEE Transactions on the Intelligent Transportation Systems*, vol. 20, pp. 4119–4133, doi:10.1109/TITS.2018.2882609

[11] R. K. Megalingam, V. Mohan, P. Leons, R. Shooja, and M. Ajay, "Smart Traffic Controller using Wireless Sensor Network for Dynamic Traffic Routing and Over Speed Detection," 2011 IEEE Global Humanitarian Technology Conference, pp. 548–553, 2011, doi:10.1109/GHTC.2011.99

[12] R. Sundar, S. Hebbar, and V. Golla, "Implementing Intelligent Traffic Control System for Congestion Control, Ambulance Clearance, and Stolen Vehicle Detection," *IEEE Sensors Journal*, vol. 15, no. 2, pp. 1109–1113, Feb. 2015, doi:10.1109/JSEN.2014.2360288

[13] C. Tang, S. Xia, C. Zhu, and X. Wei, "Phase Timing Optimization for Smart Traffic Control Based on Fog Computing," *IEEE Access*, vol. 7, pp. 84217–84228, 2019, doi:10.1109/ACCESS.2019.2925134

[14] J. Galvão, J. Sousa, J. Machado, J. Mendonça, T. Machado, and P. V. Silva, "Mechanical Design in Industry 4.0: Development of a Handling System Using a Modular Approach," *Lecture Notes in Electrical Engineering*, vol. 505, no. 3, pp. 508–514, 2019, doi:10.1007/978-3-319-91334-6_69

[15] C. Huang, R. Lu, and K. K. R. Choo, "Vehicular Fog Computing: Architecture, Use Case, and Security and Forensic Challenges," *IEEE Communications Magazine*, vol. 55, no. 11, pp. 105–111, 2017, doi:10.1109/MCOM.2017.1700322

[16] J. Feng, Z. Liu, C. Wu, and Y. Ji, "AVE: Autonomous Vehicular Edge Computing Framework with ACO-based Scheduling," *IEEE Transactions on Vehicular Technology*, vol. 66, no. 12, pp. 10660–10675, 2017, doi:10.1109/TVT.2017.2714704

[17] C. Priemer and B. Friedrich, "A Decentralized Adaptive Traffic Signal Control Using V2I Communication Data," *IEEE Conference on Intelligent Transportation Systems Proceedings, ITSC*, pp. 765–770, 2009, doi:10.1109/ITSC.2009.5309870

[18] X. Hou, Y. Li, M. Chen, D. Wu, D. Jin, and S. Chen, "Vehicular Fog Computing: A Viewpoint of Vehicles as the Infrastructures," *IEEE Transactions on Vehicular Technology*, vol. 65, no. 6, pp. 3860–3873, 2016, doi:10.1109/TVT.2016.2532863

Chapter 9

Case study on fog computing with the integration of Internet of Things

Applications, challenges, and future directions

Taskeen Zaidi and Adlin Jebakumari

CONTENTS

DOI: 10.1201/9781003230236-9

9.1 INTRODUCTION

The Internet of Things (IoT) is an autonomous system in which any device may communicate without human interaction. The IoT is an emerging area which is used in various domains. Some applications of IoT deployed in domains like industries, healthcare, home automation, energy management, and transportation and in various other areas as shown in Figure 9.1. The IoT is an emerging area; however many issues and challenges related to infrastructure, communication, protocols and standards are yet to resolve. The main aim of this chapter is to discuss a general overview of IoT, architecture, layers in IoT, services and so on.

This figure represents the applications of IoT in various industries like agriculture, healthcare, home automation, energy management, security and transportation system.

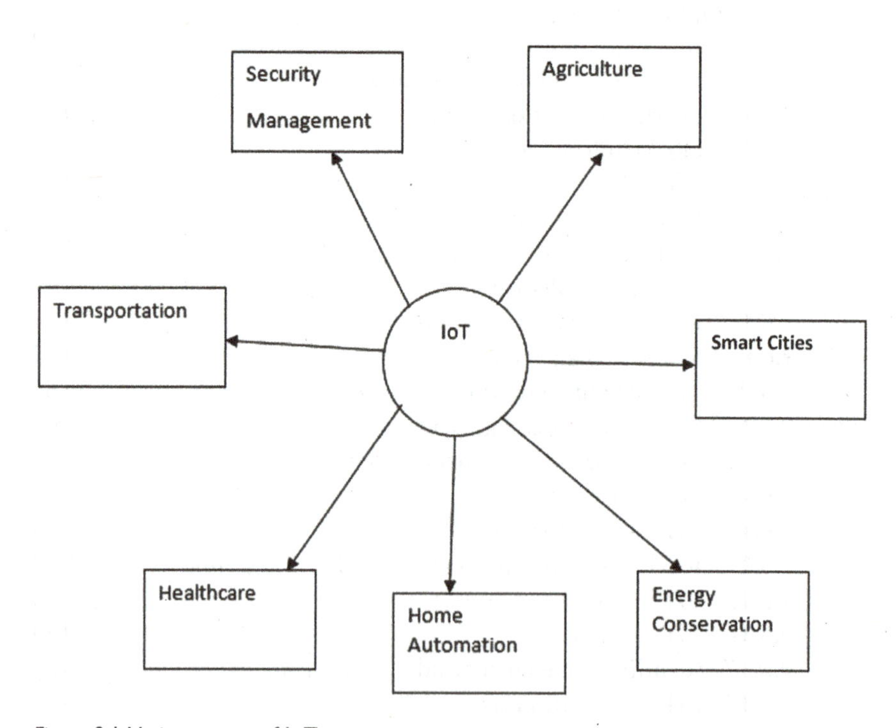

Figure 9.1 Various areas of IoT.

IoT is basically a loosely defined network of connected devices from home electronic devices, connected vehicles and sensor-enabled devices, and data is shared in a non-uniform way as represented in Figure 9.1. Due to the emergence of cloud computing or on demand-based computing which offers services to the users remotely using Internet like Amazon Web Services, the quality of service (QoS) of IoT-enabled system was inefficient as IoT-based applications are time sensitive and require security, safety, privacy and location accessibility. The two-tier architecture was not performing well for service availability. There are various challenges with the cloud infrastructure as it is difficult to connect billions of IoT-enabled system under cloud framework. Due to this shortcoming, cloud–fog architecture was proposed. The fog computing is helpful in solving the issues like network latency and traffic overloading during real-time communication. The fog computing associates cloud assets to the IoT device by expanding the services of cloud infrastructure and accessibility of IoT devices. In fog computing, little computation is performed near edge; the fog nodes handle computing loads and act as an intermediate caching system for storing intermediate computing results. The functionalities of IoT are well explained [1]. Gershenfeld [2] explained the need of IoT in the upcoming future. The IoT is providing the solutions for different applications and services storage, information retrieval and communication using electronic systems for transferring the data. The convergence of information is done for cloud, data and communication network [3]. A survey related to security and privacy issues in IoT was done by Borgohain et al. [4]. An efficient multimedia traffic security framework for IoT is proposed by Zhou and Chao [5]. The cloud computing is not able to handle IoT applications which require immediate response in an efficient way. To address this issue, fog computing is used. A region-based text aware model and fog privacy aware role-based access control model are proposed and feasibility and efficiency are also demonstrated by Dang and Hoang [6]. The innovation is playing a powerful role in changing the living standards of larger number of people and industrial Internet is a revolution in bringing new innovation into the world [7].

Fog computing was coined by Cisco, and it is similar to cloud computing and offers resources on demand providing virtualization on edge network. The resources may be accessed for storage, computation and networking to the end users. The fog computing offers heterogeneity at the large extent, and it is an extension in cloud computing framework which is capable of providing virtual computation and storage resources and services by communicating with various nodes at different levels of stack from limited devices to infinite device range. The edge computing forward applications, data and services to the edge of the network and it help in reducing congestion, cost, latency and improved security, scalability etc. Wide ranges of applications are supported by fog with multi-tenancy feature and without mutual interference. The fog interacts with the other layers using interfaces and different communication technologies. In the current scenario, IoT interfaces including fog are more heterogeneous and act as an intermediary between cloud and IoT by offering scalability and security. The IoT-based applications moved to the edge of networks. This is called fog computing. The fog computing interconnects billions of devices as well as it is suited for all the real-time applications but the main issue with fog is QoS, bill monitoring, resources provisioning, security and privacy. Babu et al. [8] conducted a survey of cloud computing and explain the QoS metrics and future direction for using fog computing in cloud environment. The fog computing reliability issues integrating cloud computing are well explained [9]. The characteristics of fog computing like low latency, location awareness, mobility, supporting large number of devices, strong real-time applications live streaming and heterogeneity are discussed [10]. Rahmani et al. [11] have proposed fog-based system architecture for health care-based IoT system to cope with challenges in healthcare for implementation of smart e-health gateway for deployment of ubiquitous health monitoring system.

9.1.1 Elements of IoT

The IoT elements are grouped into three categories:

1. Hardware: The sensors, actuators and processors are major components in IoT-enabled application.
2. Middleware: The middleware is collection of sub-layer between technological and application levels. The middleware technology hides the details of software and allows application programmer to develop an IoT-based application independently.
3. Presentation: The visualization of IoT-based application must be simple, user-friendly and platform compatible.

9.2 TECHNOLOGIES INVOLVED IN IOT AND FOG COMPUTING

9.2.1 Wireless sensor network

1. *WSN hardware*: The wireless sensor network (WSN) offers many potential applications while compared to traditional infrastructure-based network. The sensor node transmits trans-receiver, processor, power supply, interface mechanism and storage capacity, and actuators provide computational and processing power with long battery power. The WSN monitors and tracks applications. The IoT-based application is based on WSN.
2. *WSN protocol layer*: The communication among sensor node and sink plays a major role in WSN activities. The layered architecture communicates with physical world through Internet.

3. *Data aggregation*: The secure data transmission is one of the key issues in WSN as node failure effects the functionality of the network. Secure data is shared in centralized and distributed manner. Security is one of the major issues in efficient transmission of data.
4. *WSN middleware*: The middleware technology integrates cyber infrastructure with service-oriented architecture and WSN for accessing heterogeneous application in an independent manner. The platform independence is a must for developing IoT-based application.
5. *Radio-frequency identification (RFID) communication*: The radio frequency can be used for identifying billions of objects. The RFID has been used as a key technology for designing microchips in WSN. The physical objects are identified by a tag such as RFID and quick response codes.
6. *IP protocol*: The IoT-based application requires IP address for accessing every object on Internet. The IPv6 provides IP addresses to all smart objects. The IPv6 addresses are still not distributed for public use directly.
7. *Things in the web*: www is also part of IoT as it has a huge infrastructure of smart objects. The web of things uses http protocol and web 2.0, 3.0 and 4.0 as technology for communication. Virtual mall is an example of web 3.0.

The IoT includes IIoT domains causing a wide range of industries from healthcare to transportation to surveillance system. The development of robust and effective IoT system and service is a challenge. The IoT-based technologies incorporate various stakeholders and the technology-based demands of stakeholders are variable. Some of the requirements include confidentiality, integrity, authentication, access control, non-repudiation, dependability, safety and privacy. Confidentiality enables the protection of data while integrity involves verification of data, authentication involves identification of any party involved during transaction, access control ensures service availability to authorize users, non-repudiation denies action of participants, dependability offers services to the stakeholders even in the presence of errors and failures, safety provides service availability at the user end without any delay and privacy protects personal information from intruders.

9.3 CHARACTERISTICS OF FOG COMPUTING

The fog computing offers resources to the edge of network and the characteristics include:

1. It offers low latency.
2. It offers mobility.

3. It provides service over wide geographical range.
4. It provides security and privacy to the real-time sensitive data.
5. It offers heterogeneity.
6. It offers interoperability.

9.4 FOG COMPUTING PRINCIPLES

The fog computing is a substitute for cloud computing. The fog computing model extends cloud computing node to the edge of computing networks where the data are stored. The IoT connects a number of devices to the Internet and every physical thing in this world will be connected to Internet in future. This requires a lot of data processing, and existing cloud infrastructure is not able to perform the processing of data in the required time. The bandwidth is also not supported for large amount of data transmission. Due to this demerit of cloud computing, organizations are shifting toward fog computing. The traditional cloud computing is not as per the current IoT applications demand. The major issue is network latency and bandwidth.

The fog computing was coined by Cisco's manager Mr. Ginny Nichols aiming the vision for execution of applications on billions of connected devices directly at the network edge. The customers are able to develop and run applications on Cisco platform using switch and router. The fog computing provides data storage and application services to the end users. The fog computing is designed to run latency sensitive applications for supporting future Internet. The fog computing and cloud computing complement each other with their merits and demerits and play an important role in IoT. But the privacy, security and reliability of fog computing itself are research topics to be explored.

The fog computing offers services in fields like WSN, IoT, Grid Computing, Software-defined Networking and so on. The processing in fog computing is done by data hub on devices like android mobile or on router or gateways. The data is not directly sent on cloud channels, so it reduces the data processing time and fog computing performs short-term data analytics whereas cloud computing performs long-term data analytics. The cloud server performs data computation and storage using machine learning methodology. The edge devices and sensors create and compose the data as they do not have computational power. The cloud servers are far away to respond in time and to process data. The endpoints which are connecting to send and receive data over Internet should follow privacy, security and have legal implication while dealing with sensitive data and application. The fog computing may be the best option in this situation.

9.5 FOG COMPUTING ARCHITECTURE

As per Commerce [12], fog computing consists of the following elements as listed below:

1. IoT endpoints: It includes devices like sensors, gateway, edge devices and physical devices that were used to perform various functions on end devices. The IoT endpoint collects data from various devices and processes it and then transfers it. The rest data is stored on cloud server for further processing.
2. Internet protocol: The end-to-end processing is done using IP network. The data is transferred from IoT end devices to the cloud environment for further processing. The distributed intelligence is used to design an effective framework for human interaction.
3. Centralized unit: The central entity is linked to cloud environment and effectively used for storing the remaining data. The central unit stores the data for business analytics.

The IoT is a popular approach used to create sustainable environment. It also reduced carbon footprint and pollution. Few resources may be used to connect everything. Some of the popular applications which use IoT and fog computing are as follows.

9.6 APPLICATION OF IOT AND FOG COMPUTING

9.6.1 Energy efficient datacenter

The Internet is one of the widely used mechanisms for transmission of information globally. So due to vast amount of data and traffic online, the power consumption also increases rapidly. Servers were continuously active and data consumption rate is higher. But due to the emergence of data center based on IoT the energy consumption rate will be dropped. The IoT-based data center brings down the energy consumption when server loads are lower. It is an efficient approach to save money and energy concurrently. Cloud computing is on-demand computing technique that stores a lot of data and IoT may be helpful in generating and analyzing data in batch or stream format. The stream processing engines may provide fault tolerance and perform load balancing to elastic cloud environment. The IoT-based applications may be useful in offering cloud-based services and resources to fulfill on-demand computing resources and most of the IoT-based services were deployed in three-tier architecture in which bottom layer consists of IoT devices, middle layer is for cloud service provider and various applications and protocols were hosted on top layer. The important usability of IoT-based applications is to minimize end-to-end delay.

9.6.2 Home automation

The smart home is a very popular approach which uses sensors and actuation techniques, and it is reliable as a person trust that technology played a very important role to address security concerns at their home. Various sensors were connected in smart home which is helpful in providing intelligent and automated services. They help in automating customers' daily routine and also conserve electricity at home by turning off lights and electronic gadgets automatically. Motion sensors are playing an important role in resolving security concerns.

Smart homes are very beneficial application for elderly peoples and persons having disability. The sensors monitor the health of elderly person and in case of emergent situation informed immediately. The CCTV cameras are helpful in detecting the events and helpful in capturing and observing any irrelevant activities. Various sensors could be deployed inside homes or buildings to get the information about temperature, humidity, light or gas levels. The fog node may be deployed in house or building to control the waste of energy or water.

9.6.3 IoT in agriculture

The IoT-based soil moisture sensitivity system reduces excess water and ensures correct water requirement for the growth of crops. An automated sprinkler system is also used to save the crop from droughts and also used to connect sensor-based system for preventing water shortage to reduce economic crisis in agriculture industry due to drought. The temperature and humidity directly affect the agriculture productivity. The IoT-based sensors were used by farmers to measure these parameters for crop production monitoring. This application also monitors the weather.

Farming is one of the important sectors in any society or organization development. Due to inadequate absence of technology and traditional pattern, it is not easy to detect and handle the diseases in crop. Major crop damage occurs due to lack of awareness. The world is advancing in using technologies in various areas and it can be useful in agriculture sector also. The IoT sensors are able to observe the crops in the field. The IoT-based application is useful in sensing the agriculture field data as well as environmental conditions and other factors. This will be helpful in crop management.

The IoT may be helpful in crop monitoring, disease prediction, climate prediction, soil texture monitoring etc. The IoT uses drones, sensors, image processing techniques and analytical tools and machine learning algorithms for crop observations and field monitoring using sensor nodes. The data is collected and stored on cloud datacenters for further processing. The agriculture expert can guide farmers in crop protection and disease identification.

The end user can monitor the agricultural field and be able to monitor crops by taking precautions. The crop monitoring system integrates IoT-fog and machine learning. The data collection, analysis, filtering and alert detection are done. The decision-making is applied on local edge network and related information will be communicated to the farmer about any alert situations. A study based on acetylcholinesterase (AChE) biosensor and IoT was done by Zhao et al. [13] for detecting pesticide residues in agricultural field.

9.6.4 Reducing pollution

The IoT-based sensors help in sending and receiving data from vehicles and other sources to reduce pollution and energy consumption. A real-time alert will be generated by the system to inform about the precautions accordingly. Air pollution can also be monitored by IoT-based applications. The vehicles which are prone to severe pollution may be identified by RFID tags.

9.6.5 Smart transport system

This application manages traffic in cities. The sensors and intelligent processing systems were used to monitor transport daily. The objective is to minimize the traffic congestion, monitor parking system and avoid road accidents by monitoring and controlling vehicle movements. The GPS and gyroscopes were used for keeping a check on passengers and vehicles and cameras were used for monitoring vehicles and traffic. The IoT can be used to connect public transport system. The self-driving cars launched by companies like Tesla are real examples of this approach. The IoT-based system reduces pollution and optimizes the movement of people and services. The industrial IoT-based system has impacted tracking, traffic management and in many services in a broader way.

9.6.6 Traffic surveillance

Sensors were deployed to monitor traffic in various parts of the city. Traffic congestion can also be detected using mobile phones and GPS sensors. The smart connected vehicle systems are located inside vehicle for controlling sensors and actuators in the vehicle, like tires temperature, pressure and street lane. The information is collected by different sensors and sent to fog nodes and the real-time responses regarding dangerous situations will be informed.

9.6.7 Intelligent parking monitoring system

The free parking slots will be checked on Internet to find out parking space. The free slots availability will be monitored by central server.

9.6.8 Safety management

Many IoT-based applications were deployed, which will be helpful for drivers to monitor safe driving. The driver behaviors can be also observed through smartphone sensors. The accelerometer and acoustic data will be helpful in detecting accidents. This IoT-based application can be installed on smartphone.

9.6.9 Traffic lights monitoring

The traffic light sensors may be useful in sensing the congestion and traffic status ongoing. The traffic lights also give a pass to emergency vehicles if any situation occurs. The smart traffic light handles congestion in network by detecting vehicles, roadside components and identifying connectivity of bikes, vehicles or ambulances on the road. The traffic lights are fog devices which turn green or red depending on congestion in one direction. The smart light also detects emergency vehicles.

9.6.10 Smart water management

Due to water scarcity problem all over the world, it is very important to manage water resources in an efficient manner. A smart solution is to place a lot of meters in water supply lines. The meters will be capable of measuring the inflow and outflow of water and detecting water leakage, if any.

9.6.11 Health care management system

Many IoT-based applications were available to monitor personal health condition of a person. The IoT-based sensors were used to monitor and record health updates and to monitor if any abnormal condition arises. The IoT-based applications also record all medical details of an individual like allergy, blood pressure and sugar level. The IoT-based stress identification applications are also very popular. This application can be used on smartphone. The potential market value of health care system is estimated up to $2,737 billion by 2022. The Internet of Medical Things (IoMT)-based applications utilized user equipments to acquire data on the IoMT servers from the patients anywhere. The fog computing system will be helpful in managing patient's records in a better way by offering reliability and privacy to the patient data. Mason [14] proposed that the fastest-growing M2M sectors offer security to healthcare and other services. Sudhakar Yadav et al. [15] have proposed a framework on IoT for tracking the status of medical records, detection of chronic disease, prediction of Parkinson's disease, health monitoring and management.

Reducing network latency is one of the benefits of using fog computing as compared to cloud. The fog computing also offers privacy to the user data as compared to cloud. It analyses the data on local fog node instead of sending on cloud data center. The fog nodes are geographically distributed, so one fog node is able to handle the request of one area. This reduces the data to be sent directly to the cloud data center for analysis and processing. Due to reduction in data on cloud data center, the network bandwidth can be properly utilized. The fog computing also offers energy efficiency and scalability.

It is a big challenge to handle data in health care systems due to increase in number of patients and diseases. Most of the hospitals are capturing patient records manually. It is very difficult for nurses and hospital to record the health data as this activity consumes time and cost. The automatic process of analyzing the patient health data will reduce the time and cost. The integration of fog computing in healthcare enables the remote monitoring of patients, and data is generated using a sensors network. The sensor node is transmitting the data on fog node which is accessible to clinical staff.

9.6.12 Forest fire detection

Smart UAVs are inexpensive and can be useful to monitor large areas under different climatic conditions during day and night for detecting and controlling forest fire. The UAVs perform real-time event reporting using image-based identification program. The Internet connectivity helps in monitoring the forest area.

9.6.13 IoT in data analytics

The IoT is useful in creating a ubiquitous computing environment which processes a huge data and is stored in distributed or heterogeneous manner. As big data is characterized by velocity, volume and variety, and different data processing approaches were used for data analysis. The IoT-based applications were used for data collection and analysis. The pattern detection and data mining techniques were used for data extraction and knowledge discovery. The ellipsoidal neighborhood factor may be implemented for configuring different host devices. Due to emergence of new mobile-based applications like augmented reality and virtual reality, more computational power will be delivered by fog nodes for effectively distributing the process to all the connected nodes.

Various companies and organizations capture data from sources like social networking sites, emails, online surveys, web server and sensors. The data may be structured, semi-structured and unstructured and coined as big data. Most of the organizations apply analytics for data analysis to discover hidden patterns, correlations and other useful information. Some data are stored and processed in real-time mode and require real-time processing. The big data analytics improves business decisions, reduces cost, analyzes information for better decision-making and offers new solutions and services to the customer. An analytics engine was proposed [16] to get sensor data from various devices for analyzing the information from IoT data using machine learning algorithms.

9.6.14 IoT-based real-time analytics in fog computing

Data analysis will be done on big data where data may be generated from sensors and devices in real-time applications. The real-time processing and requirement of high-performance computing power lead to the development of edge computing platform. The edge layer devices are closed to end user and devices were connected to edge cloud which further decreases the network delay, improve energy consumption, save processing power, and provide fault tolerance computing.

The fog computing is also an extension of cloud computing that offers features similar to cloud-like network bandwidth, Internet characteristics, storage capacity, low latency and efficient service level agreement and maintains QoS requirements.

Several virtual reality solutions were used like oculus rift, Google dashboard and so on in wired and wireless environment. The data processing is done on fog nodes which offer better performance and conserve energy.

9.6.15 IoT in business process

The big data analytics may be useful for fields like industry, organization and an individual to improve efficiency and productivity of an application.

9.6.16 IoT security

Due to the increase of IoT-based application users, the attackers are targeting end devices. They want to access the smart devices and sensors and attack the devices which are directly connected to end users, so valuable information may be removed or omitted. As IoT-based techniques are distributed in nature it is easier for intruders to harm the end devices. But cryptographic approaches may end-to-end encryption by maintaining the confidentiality and integrity of user data. Various knowledge discovery-based algorithms were designed to protect devices from vulnerable attacks.

The IoT-based techniques were used to fetch information related to community online searching habits by observing real-world events automatically. The smartphones are connected to number of sensors and actuators. In today's scenario, cars have also sensing capability which will be helpful in providing information about car movement as well as road conditions and traffic scenario. The Intel has also designed an intelligent systems framework for managing and securing devices of IoT configured data.

9.6.17 Identity management and privacy using IoT

Devices connected in IoT are identified by ucode and can be used RFID tags to create uniform resource identifiers to locate objects and to decrease the complexity of running devices connected through IoT-based applications. The automatic authentication and context pairing can also be used with IoT devices.

As data is increased day by day in abundant amount, the security of data is an important concern as well as security mechanism is also required while sharing and storing data.

9.6.18 IoT in education sector

The interactive learning platform is very helpful and popular in today's education offering system as we can add study materials in the form of texts, audio, video, animations, assessments, quizzes etc. to make it an effective way of learning and sharing knowledge.

The students were able to gather the subject knowledge in a better way and interact with friends and teachers in an easier way. The real-world problems were discussed in an effective way and students were able to find numerous ways to resolve the problems through discussion. Many students were present in class, sometimes monitoring students; presence and performance is risky. A smart security is required to avoid the risk factors. The IoT-based applications may be an effective way to enhance the security approach during interactive teaching and learning. The student activities in a class may be monitored by 3D positioning technologies. The intelligent cameras may also install in an organization to keep a track on any illegitimate activity.

The IoT-based mobile apps provide a very powerful technology. The students and teachers will be able to create multimedia-based graphics. The information can be presented in an effective manner like creating video demo or animation-based games. This new feature is offering more opportunities for an effective teaching and learning process. The IoT-based attendance monitoring system is also very impactful as students' daily attendance record will be shared automatically with central server without any human intervention. This will be helpful in concentrating more on teaching and learning and proper utilization of class timing. Fog computing techniques can be used to maintain the privacy of student record by avoiding uploading and sharing of sensitive data of their grades on cloud-based systems.

9.6.19 IoT role in oil and gas industry

IoT can be used in oil and gas industry for process automation. The IoT-based technologies can be used for service applications, accessing wireless fields, remote monitoring of oil and gas plant. For the purpose of monitoring, the field device is developed and installed in areas under tough environmental conditions.

9.7 ADVANTAGES OF USING FOG COMPUTING FOR IOT

1. *Minimizing latency*: The raw data were analyzed by edge nodes quickly, so it reduces the latency of the control loop.
2. *Improving reliability*: The local processing is done to manage the breakout in the cloud framework. The data centers offer storage, processing infrastructure, network connection and reliability.
3. *Improving privacy and security*: The IoT data is sensitive and requires proper storage. The local gateway or edge nodes can be used to control the storage providing security to data.
4. *Bandwidth proper utilization*: This framework offers data processing locally, and IoT-based systems were connected to cloud to limit the intermediate connections preserving wastage of bandwidth.

9.8 CHALLENGES FACED BY FOG ARCHITECTURE

Fog computing is a promising area to boost IoT-based applications. But still there are many challenges in fog model setting. The deployment of multi-level structure consisting of group and subgroup of nodes in a hierarchical framework would provide elasticity and scalability to real-time sensitive communication. The fog setting must be done to optimize the systems performance in vehicular applications. Fog node may be composed of specific and concentrated resources to perform the task assigned. The actuation capacity must be deployed for evaluating the traffic pattern for identifying the ways to reach the destination. The fog framework should be designed for cloud to optimize the communication between fog and cloud components. Specific interface is required for latency tolerance application and latency sensitive application. The efficient data processing mechanism is required for handling large amount of sensor real-time data. Internetworking of different fog nodes to aggregate data from sensors located in various networks and also sharing of data in different networks is also necessary for offering services in a distributed manner in IoT-based applications.

9.9 CONCLUSION

The IoT is one of the fast-growing technologies in today's perspective and in the current chapter a survey on IoT-based technologies was discussed. It was concluded that IoT has been emerged as a new paradigm which provides energy consumption, data analysis, communication etc. The IoT-based applications can provide energy conservation, transmission, energy efficiency and increasing sources for renewable sources of energy in various domains of information technologies with reducing environmental factors. The IoT-based technologies provide various applications and platforms for data analysis. The privacy and security issues related to IoT and its impact on future were also discussed in the current chapter. The fog computing is a promising technology to enhance the performance of IoT-enabled system. The role of fog computing in industries is needed to be established practically for deploying sensors-based real-time sensitive applications.

REFERENCES

[1] Casaleggio Associati, "The Evolution of Internet of Things," February 2011, online at www.casaleggio.it/pubblicazioni/Focus_internet_of_things_v1.81%20-%20 eng.pdf
[2] N. Gershenfeld, "When Things Start to Think," Holt Paperbacks, New York, 2000.
[3] World Economic Forum, "The Global Information Technology Report 2012—Living in a Hyperconnected World" online at www3.weforum.org/docs/Global_ IT_Report_2012.pdf
[4] T. Borgohain, U. Kumar, and S. Sanyal. "Survey of Security and Privacy Issues of Internet of Things." arXiv Prepr. arXiv1501.02211; 2015.
[5] L. Zhou L and H-C Chao. "Multimedia Traffic Security Architecture for the Internet of Things," *IEEE Network*, vol. 25, no. 3, pp. 35–40, 2011.
[6] T. D. Dang and D. Hoang, "A Data Protection Model for Fog Computing." In: Fog and Mobile Edge Computing (FMEC), 2017 Second International Conference on IEEE. IEEE, pp. 32–38, 2017.
[7] P. C. Evans and M. Annunziata, Industrial Internet: Pushing the Boundaries of Minds and Machines, General Electric Co., online at http://files.gereports.com/ wp-content/uploads/2012/11/ge-industrial-internet-vision-paper.pdf
[8] R. Babu, K. Jayashree and R. Abirami, "Fog Computing QoS Review and Open Challenges," *International Journal of Fog Computing*, vol. 1, no. 2, pp. 109–118, 2018.
[9] H. Madsen, G. Albeanu, B. Burtschy, and F. L. Popentiu-Vladicescu, "Reliability in the Utility Computing Era: Towards Reliable Fog Computing." In: Systems, Signals and Image Processing (IWSSIP), 20th International Conference on IEEE. IEEE, pp. 43–46, 2013.
[10] F. Bonomi, R. Milito, J. Zhu, and S. Addepalli, "Fog Computing and Its Role on the Internet of Things." In: Proceedings of the First Edition of the MCC Workshop on Mobile Cloud Computing. ACM, pp. 13–16, 2012.
[11] A. M. Rahmani, T. N. Gia, B. Negash, A. Anzanpour, I. Azimi, M. Jiang, and P. Liljeberg, "Exploiting Smart E-health Gateways at the Edge of Healthcare Internet of Things: A Fog Computing Approach," *Future Generation Computer Systems*, 2017. www.sciencedirect.com/science/article/ pii/S0167739X17302121
[12] Mind Commerce, "Computing at the Edge Series: Fog Computing and Data Management," 2017, www.mindcommerce.com/files/FogComputingDataManagement.pdf

[13] J. G. Zhao, Y. Guo, X. Sun, and X. Wang, "A System for Pesticide Residues Detection and Agricultural Products Traceability Based on Acetylcholinesterase Biosensor and Internet of Things," *International Journal of Electrochemical Science*, vol. 10, no. 4, pp. 3387–3399, 2015.

[14] Analysys Mason, "Imagine an M2M World with 2.1 billion Connected Things," online at www.analysysmason.com/about-us/news/insight/M2M_forecast_Jan2011/

[15] N. Sudhakar Yadav, K. G. Srinivasa, and B. Eswara Reddy, "An IoT-Based Framework for Health Monitoring Systems: A Case Study Approach," *International Journal of Fog Computing*, vol. 2, no. 1, pp. 43–60, 2019.

[16] K. Ashton, "That 'Internet of Things' Thing," online at www.rfidjournal.com/ article/view/4986, June 2009.

Chapter 10

Virtual health management through IoT

Anushka Phougat and Anil Saroliya

CONTENTS

DOI: 10.1201/9781003230236-10

10.1 INTRODUCTION

The Internet of Things (IoT) is defined as a network of billions of devices that can observe, interact, and exchange data, which can then be analyzed to reveal a wealth of practical insight into planning, management, and decision-making [1].

IoT not only increased independence but also distanced humans' capacity to engage with the outside world. IoT, with the assistance of future protocols and algorithms, has made important contributions to worldwide communication. It links to the Internet a huge number of gadgets, wireless sensors, household appliances, and technology devices [4]. IoT is being used in the agriculture sector [5], automotive, residential, and health care.

The growing popularity of IoT is due to its advantage of demonstrating high accuracy, low cost, and the ability to predict future events in a better way. In addition, increased software and application knowledge, with the development of mobile and computer technologies, easy access to wireless technology, and the growing digital economy add to the rapid transformation of IoT.

IoT promises major benefits for various sectors such as health care, production, agriculture, telecommunications, and transport. Aside from the IoT concept's popularity and promised benefits, its adoption is much slower than expected. Some of the key causes behind this are as follows:

1. Security, privacy, policy issues, and trust.
2. Organizational inertia, long-term financial cycles, and shortage of professional staff needed to effectively use IoT.
3. The lack of convincing use cases with a clear refund investment (ROI) in other sectors.

This chapter bestows a high-level view of the technology involved in the helpful research on *IoT and Cloud Computing*.

COVID-19 has had a sincere effect on all realms of life, so much so that it is unlikely that we will ever return to normal. This epidemic seems to be the basis for digital transformation because COVID-19 has created or expanded systems and is using digital technology cases [4].

The key to improving social and moral conditions is to change the priorities and views of governments, organizations, and individuals. In many cases, this has been the case to resolve or reduce many of the above causes caused by the expected adoption of IoT on multiple verticals. For example, governments have invested heavily in IoT resources and other technologies to combat COVID-19. Lifestyle changes introduced by COVID-19 such as working/studying from home have also provided new IoT usage cases with a clear ROI such as remote asset management, staff tracking, and the interaction of remote workers [2]. As a result, the number of organizations has increased investment in IoT and the speed of their IoT projects [3,4].

Anti-COVID-19 has led to a strong allegiance to intimacy issues, high technology confidence, and speedy authorization processes. This also paves the way for faster IoT adoption in various verticals. In addition, control changes such as strict cleaning and tracking of business needs also accelerate the adoption of IoT in smart devices. In this paper, we discuss the potential impact of COVID-19 on IoT in different sectors, namely health care, smart housing, smart buildings, smart cities, transport, and industrial IoT.

We found that COVID-19 did not positively affect IoT adoption in all fields, at least in the short term. For example, mainly due to the economic downturn, the majority of car companies are not in a position to invest in IoT systems. As a result, this section discusses the short- and medium-term effects of COVID-19 on IoT adoption in these fields. In addition, for each category, we also discuss those new programs.

These are considered the challenges that need to be considered, and the important indicators of research that will help IoT adoption. This is the first activity that explains the impact of COVID-19 on the adoption of IoT in various fields. However, our work is very different from the following major features [3]. First, the excess most existing jobs focus exclusively on IoT in the health care sector. This is not surprising because COVID-19 is a health problem.

However, COVID-19 affected almost all areas of our lives and we provide an initial report on its impact on IoT in various forms and key areas including health care. Second, existing research focuses primarily on how IoT and other digital technologies can be used (or are being used) to combat COVID-19.

10.2 IOT AND ITS BACKGROUND FOR COVID-19 PANDEMIC

The present worldwide epidemic of COVID-19 has transcended provincial, strong, intellectual, spiritual, social, and educational obstacles. Patient satisfaction and fewer hospitalizations using the IoT to effectively cut healthcare expenses and enhance treatment outcomes for infected patients. Finally, the current research-based study aimed to identify, explain, and underline the full execution of the existing IoT concept by providing a pathway for recognizing the COVID-19 epidemic. Finally, 12 significant IoT applications were identified and explored. Finally, researchers, intellectuals, and scientists were compelled to devise effective strategies to overcome or battle the epidemic known as the fourth industrial revolution. Industry 4.0 was capable of responding to individual needs during the COVID-19 tragedy.

The transformation begins with the use of advanced manufacturing technologies and digital, technical information. Many useful Industry 4.0 technologies are helping to effectively manage the COVID-19 epidemic, and this will be discussed in this document. The available Industry 4.0 technology can help detect and test COVID-19 and other related symptoms and problems [13].

Simply put, the IoT is a system of connected devices/functions that are compatible with all networked entities such as hardware, software, network connection, or any other necessary electronic/computer equipment that ultimately enables them to respond to disputes and collect data through efficient integration and exchange between consumers and service providers. In the current general situation, most of the problems arise from poor access to patients, which is the second most important problem after fears about the development of a vaccine [3]. Using the concept of the IoT makes patient discovery very useful, ultimately helping them provide the care they need so they can recover from their illness.

10.3 CLOUD COMPUTING ASSISTANCE IN COVID-19

The wireless infrastructure developed appliance may compile data for a longer duration merely with a tiny strength authority as they accomplish unique methods of an anthology of health-related data and transmission to gateways, the sensing data of healthcare monitoring consumes minimal energy but they have limited analysis power to filter this data when cloud computing plays a critical role and complements the gap of wireless infrastructure established system in cloud computing with its tremendous computation energy for the easy deployment of healthcare monitoring algorithm and assists to refine sensor data [6].

The healthcare industry has remained dormant in terms of cloud computing technologies. Nonetheless, recent changes in circumstances and advancements over the years have pushed them to shift from data-based solutions to cloud-based solutions.

The cloud computing platform aids in the production, management, and distribution of open-source pharmaceuticals. Cloud computing technology provides researchers and scientists with the accessibility and flexibility they need to comprehend data and information and create prospective ideas quickly and efficiently.

While one does not often consider the linkages between the biological business and cloud computing, it is astounding that this is the case. Cloud computing has played an important role in the advancement of vaccination. Its relevance was heightened during the COVID period, when hundreds of vaccination companies, researchers, scientists, and government organizations expressed their belief in the persuasiveness and benefit of cloud computing in the suggested solution that communicates with the deadly virus.

10.4 BENEFITS OF CLOUD COMPUTING

10.4.1 Diversity

Given the need for coronavirus vaccinations, which intensify the pandemic, hospitals, and labs have never been able to rely on the current IT infrastructure. Due to a lack of hardware and CPUs, the built-in IT infrastructure will be unable to correct massive amounts of data. Additional hardware is required to control the flow of data, and adding new hardware and applications may be highly beneficial [7].

The use of cloud technology has enabled health care businesses to increase their processing and storage capacity without the need for additional hardware. Cloud computing simplifies healthcare technology such as electronic medical catalogs, mobile applications, patient portals, IoT gadgets, and big data analytics. It provides a paucity of complexity and flexibility [7].

10.4.2 Processing power

Cloud-based solutions can update and improve their characteristics at a rapid pace with the smallest actions, and you can receive real-time updates and any relevant data. Biomedical research involves massive data sets and numerous variances. High-performance computing (HPC) manages the acquisition of numerical data. HPC computers enable machine learning methods and the rapid processing of large amounts of data.

The terrible issue is that most academics do not have access to these tools, causing policy-making to stall.

In such cases, cloud technology looks to be a lifesaver. It has enabled organizations to rapidly increase their processing capacity by utilizing cloud apps while eliminating much-needed hardware costs.

These future technologies have also altered the scope of clinical research, and the cloud may be used to manage clinical trials and exchange data [7].

10.4.3 Data sharing

The data-sharing process has become considerably simpler and more controllable thanks to cloud computing. Healthcare organizations have successfully used the cloud to exchange and send data to a centralized channel to speed up the vaccination process and reduce transmission length.

Because healthcare information wants to remain private and in the cloud, it may be safely handled in real time by all relevant healthcare stakeholders such as physicians, nurses, and carers.

The cloud enables users to have access to a common repository of analyzed data and test results, allowing them to design the system and reduce settings and faults [7].

10.4.4 Expenses

The cloud can store large amounts of data at a very low cost. Cloud computing is based on a pay-as-you-go and subscription model, which means you only have to pay for the services you utilize. It eventually allows small-budget minor sanitariums to use a cloud-based prototype [7].

The present increase in COVID-19 disease has affected many aspects of our daily lives. In the meantime, virologists are scrambling for new vaccine therapies.

10.5 PROCESSES INVOLVED IN IOT FOR COVID-19

Examining cutting-edge and out-of-the-box technical explanations for combating the COVID-19 pestilence through the integration of innovative IoT technologies, such therapies could result from the output of IoT-based industrial ventilation, masks, and additional medical tools used to survey patient conditions in sanitariums or isolation, as well as new communication, isolation, and IoT-based COVID-19 testing strategies, as well as mobile data collection and analytics, in the advancement of smart point-of-care not limited to the modern gold standard of reverse transcription chain polymerase reaction. To begin, IoT adoption frequently offers hurdles in terms of connection, power, spectrum, and bandwidth needs, as well as cost. However, falling computer costs (including sensors) and increased mobile broadband penetration will accelerate the application of the IoT in healthcare. The low cost of wireless technology *with low power consumption will* also *contribute* to this trend.

A large number of health problems require a physical examination to be diagnosed. In addition, images and videos transmitted by telemedicine using IoT technology can be in high resolution, making healthcare essential [8].

Technology can improve sympathetic maintenance by connecting patients with healthcare providers and facilitating the gathering and analysis of data that might improve patient care. Here are some of the most important IoT-enabled applications in developing areas.

10 5.1 Convenience

It can supervise to improve accomplishment for beginners: by exceeding inward authorizations without reducing physician and sympathetic time, strategies can switch to an automated planning system with close, rapid communication abilities provided remotely.

10.5.2 Timeless

Another important feature of IoT therapy is the ability to regulate previous treatments. Particularly in severe, critical medical circumstances, the additional minutes provided by continuous monitoring might be the difference between life and death.

10.5.3 Accurateness

Another important component of IoT health care is predictive analytics, which may improve efficiency and prompt action. Attaching all devices and alarms to the same complex GPS allows not only for the early intervention of noted patient circumstances but also for support squads to respond swiftly and confidently to crises, which is especially advantageous for patients living in rural or distant locations.

10.5.4 Safety and insurance

Patient monitoring not only permits greater physical protection with better fall prevention appliances but also increases digital defense by accessing a sufficient secure network for obtaining patient health data.

10.5.5 Mobility

It can improve the patient's dignity sufferers are increasingly pleading for access to care for their everyday consumer equipment, a trend that has been well noticed with the rise of FitBit.

10.6 IOT FOR VERIFICATION AND TRACKING

IoT testing and COVID-19 compliance can help to prevent disease transmission since they are important in the battle against the disease. As a result, the IoT is now commonly used to test and evaluate apps that expedite the IoT.

The IoT is employed in medical, industry, transportation, and the environment, and it has several advantages over non-Internet-connected technologies. In the case of COVID-19, it is particularly effective as a preventative strategy, allowing for early illness discovery, resulting in less therapy and better treatment, control, improved diagnostic accuracy, low cost, and lower chance of mistake 8. These advantages must be balanced against the technological challenges. IoT systems are vulnerable to security and privacy risks [9].

Cyberattacks are predicated on a lack of authorization and authentication, a lack of interface protection, or a lack of encryption.

A wide number of nations and regions, including the United Kingdom, South Korea, Germany, Spain, Vietnam, and Taiwan, have employed digital verification and tracking. While some of these attempts were unsuccessful in locating COVID-19, governments utilized this technology extensively in their remedies, and the disease was handled considerably better.

Internet-connected tools make it easy to disseminate and collect the data needed to make informed decisions without significant human intervention [10]. IoT systems are composed of five layers, which comprise embedded systems such as processors, sensors, and communications in smart web gadgets that enable data gathering and sharing. The knowledge, network, middleware, application, and business layers are among them. Mobile devices such as sensors or barcodes that integrate information provided to the network layer constitute the lowest level of vision.

The network coverage area enables data to be transported from the visual level to the data processing system via wireless or wireless communications such as mobile standards (4G, 5G), Bluetooth, or Wi-Fi. The following layer of middleware will modify the acquired data and make the appropriate judgments depending on the findings obtained from computing devices all over the world.

10.7 PROBLEMS OF HANDLING IN TRACKING FOR COVID-19

These challenges can be mitigated through the use of digital tools *embedded* in public health systems. *Digital tools based on applications can aid in the contact tracing process.* Some of these digital tools, such as *explosion* response, signal tracking, and proximity *tracking*, have been used to track a *contact.*

Contact tracing has been difficult due to:

1. Inadequate contact identification,
2. Poor reporting systems,
3. Complex data management requirements, and
4. Process delays, increasing the time from contact discovery and separation of suspicious cases. In addition, this human-centered approach also has serious data security concerns.

Wearable technology is currently being used not just to combat COVID-19, but also in resilience and health, and has become increasingly acceptable to the general public even outside of those fields.

Following the creation of these wearable gadgets, necessary, dependable, and representative contact information is frequently gathered from the general population. Furthermore, when these gadgets rely solely on Bluetooth, except for GPS monitoring, people are more inclined to utilize them because the capacity to violate privacy is limited. As a result, the ideal contact tracking tool will be a wearable item used in conjunction with an appropriate digital tool, such as a smartphone app. The tremendous knowledge problem posed by wearable gadgets and phone apps might be transformed into analytical opportunities.

10.8 NEXT-GENERATION PROPOSED TOOLS AND DISCUSSIONS

In recent years, the IoT is becoming more and more in style as it relates to health monitoring and surveillance. A patient infected with an infection may also be tracked with IoT throughout the illness. IoT-based COVID-19 and alternative communicable disease chase models were developed. The improved model suggests a digital contact tracking system to trace a COVID-19 infected patient. The advanced model also incorporates blockchain technology to track infected patients' victimization through mobile signal technology. In [9], new IoT technologies are used as a monitoring system during the extinction epidemic. Advanced IoT technology is being developed in such the simplest way that it is often activated throughout a virulent disease and has methods to contend with the epidemic and supply processed and versatile treatment during the epidemic. The primary epidemic observance framework and IoT-based response framework are developed. Associate degree improved IoT framework that identifies COVID-19 within the first place and helps in period monitoring to trace down patients with the disease in the event they receive treatment [11].

10.8.1 Deep machine learning

The study conjointly explores the use of digital technology in medical transformation and also explores the way to curb the unfolding of the COVID-19 epidemic.

We are now living in an era of big data, where scientific and technological research generates huge amounts of data. Given the various challenges in analyzing and interpreting big data, it is necessary to move from traditional methods to more advanced methods of data analysis. One of the more advanced is advanced learning, a new concept of artificial intelligence (AI). In theory, deep learning is a subset of machine learning. However, unlike traditional methods, deep learning models are composed of many layers of deep communication capable of extracting information from large amounts of data. This may avoid functional engineering, which is a data reduction strategy. As a result, deep learning algorithms are being developed in many areas of public health, from disease diagnosis to mass surveillance. Other widely used deep learning models include physical signal analysis and cumulative neural networks, autoencoders, deep belief networks, short-term memory networks, and deep feedback neural networks. AI, including standard machine learning, and extended models are also used to manage

COVID-19. In order to complete and validate educational activities quickly, deep learning models require a lot of processing power.

The use of *machine learning* IoT to *track outbreaks*. The devices listed were used to scan the default location.

At the next stage, the sensor reads the temperature of the suspect entering the restricted zone from the unrestricted zone. If the temperature exceeds 98.6°F, the suspect will be identified as a patient. Otherwise, detailed information for the future will be collected and presented, and the suspect can leave. If possible, when a suspect is found to be infected, he will be treated as a patient. Your location information and mobile phone GPS data will be downloaded. The collected data will be stored on a cloud server and can be shared with interested people and government organization for treatment. Infected patients will be monitored and their movements monitored to prevent the spread of the epidemic. In case of nonobservance of the rules of the protected area, the suspect may be pursued and disqualified. This contributes to the further spread of the disease.

ML-based is a low-power, efficient, and easy-to-use system that defines a system for any predefined problem. Sensors are real-world data providers that are already outside the organization, and actuators allow objects to act or react like information received from sensors. To prevent diseases from spreading, an IoT project is proposed.

The data connection associated with the utility's functionality will be redirected to the cloud corridor. In a large data distribution center, classification of information, that is, complete information, is retrieved. Having a large data center is a result of organizing information. ML-based AI is used to create framework-based models and get data. Knowledge tests can be used to present results and optimize performance. Infrared sensors can be used in open bathrooms for convenient scheduled access to water. Infrared thermometers can be used to measure indoor temperatures to detect contamination between pools and surface gaps with optical cameras for aviation departments, train stations, tourism, shopping malls, etc.

Sensors offered in the engineering field can be used to monitor indoor temperature quality, planned entry and exit operations, control circuits in public places and bathrooms, and online meetings to maintain a strategic distance from personal contact with a person. In-depth exploration of intelligence can help understand health patterns, model communication risk, and determine outcomes. For personal or small applications, temperature sensor configuration, NodeMCU board, or Arduino board with sensors and Internet can be used.

10.8.2 New generation challenges

The IoT is used in medicine, industry, transportation, and the environment and has many advantages over non-Internet systems. In the case of

COVID-19, it is very useful as a monitoring system, allowing early detection of diseases, resulting in less treatment, better control of distribution, higher diagnostic accuracy, average cost, and less risk of error. These advantages must be weighed against the problems associated with technical difficulties, authenticated, visually secure web interface, or no encryption. Another problem arises from the potential for disruption to other services and/or the availability of resources due to varying data transmission rates. Time and power consumption can be disrupted on IoT devices, which creates another problem.

The biggest challenge in making contacts is understanding the dynamic nature of the epidemic. A complete estimate is not possible because the available data does not include spatial and temporal sample sizes. However, there can be mixed interpretations of the data tracking data based on information obtained through consultations with experts. While it is not possible to fully imagine sampling space, it is generally believed that large amounts of data improve data availability and improve decision support for contact tracing applications.

The amount of data far exceeds the capabilities of one person's perception; therefore, expert guidance is only possible at a methodological level, when expert groups work together on methods of data translation. Methods have been developed, and the next challenge is to develop training systems that encompass the skills that support these methods and the knowledge associated with road construction.

10.8.3 Advantages

a. The tools available are very useful for tracking a contact. The recommended tool is probably the only one useful for contact tracing and forecasting COVID-19 collection.
b. A wide range of demographic data will be available for in-depth model training.
c. Security and privacy issues will be addressed.

10.8.4 Disadvantages

a. Many countries currently experiencing financial difficulties may not be able to provide their citizens with such a tool to communicate with them.

10.9 KEY IOT TOOLS TO FIGHT COVID-19 PANDEMIC

IT is a new technology that ensures the quarantine of all those infected with the virus. The monitor will be able to keep track of blood pressure, heart rate, and blood sugar during separation.

With the successful implementation of this technology, we can see an improvement in the efficiency of healthcare workers by reducing their workload; the same can be said in the case of the COVID-19 epidemic with minimal cost and error.

Telehealth consultation: The exposure of the virus to the virus prompted doctors to examine patients via video chat to determine if the patient had been infected without contact. Electronic communication and home confinement are some of the best ways to combat the virus seen in hospitals and nursing homes.

Digital diagnostics: Various types of IoT devices are used to monitor health data after performing digital diagnostics. The advent of Kinsa smart thermometers contrasts with traditional thermometers, which can collect valuable data to share with healthcare providers and track clues to better protect the community.

Remote monitoring: It monitors chronic illnesses in elderly patients that increase the risk of dying from a deadly virus.

Robot assistance: The use of IoT robots is a growing trend. They are used for disinfecting equipment, cleaning hospitals, and delivering drugs, giving medical staff more time to treat their patients. A Danish company to keep healthcare facilities clean during a natural disaster. These robots use the IoT to help kill germs in nursing home healthcare facilities and clean up patient rooms.

Work from home and IoT: The IoT is made up of four parts: sensors, networks, clouds, and applications. The outbreak has further accelerated the adoption and possible implementation of Emergency Home Care (WFH). The teleworking trend has flared up in the aftermath of the pandemic, and it's likely to continue. The best part: it provides flexibility and less time for tasks. IoT sensors and the networks behind them have made this job more attractive and ideal for business.

Blockchain and IoT: Using blockchain, people can exchange any information/transactions in real time between relevant groups, available as secure and random websites. This is the time when all countries are fighting the COVID-19 pandemic. The past may have helped save the world from such death and suffering. This is the part where the blockchain and the IoT interact with node/sensor information and move through the networks available for work in the cloud, which will be presented at the request of the authorities and medical personnel. In short, data can be completely protected with blockchain [12].

10.10 WEARABLE IOT DEVICES

The wearable system is a smart and intelligent assistive device that is worn on the left or right ankle to transmit, test, monitor, and verify important indicators of COVID infection. Nineteen data from human sensory sensors may be classified into numerous categories. In general, COVID-19 symptoms may be separated into two categories: immune system signaling systems and other symptoms such as the respiratory system. Temperature, heart rate, and oxygen saturation are the primary indications of health in the early phases (SpO_2). Cough scoring methods are used to assess cough frequency and identify shortness of breath.

Hence, all of these symptoms need to be considered when designing a wearable device as they can play an important role in detecting COVID-19 symptoms as it is not possible to screen and monitor a single person. Public diagnostics have made it possible to diagnose the same large number of people. Let's take a look at each sensor that is used to identify and detect symptoms of a COVID-19 patient.

This layer is in charge of gathering two sorts of data: GPS sensor data area position and medical data for COVID-19 parameters such as temperature, heart rate, total oxygen (SpO_2), and cough statistics. Depending on the application's architecture, a microcontroller at this level receives data from sensor modules linked to the IoT core. The IoT program collects all data on various patient locations using a GPS sensor, determines the patient's location in real time, and saves the information in the cloud. This layer's goal is to detect atypical patient symptoms and gather GPS data to help anticipate the patient's status in one location and monitor them over time.

10.11 DRONES TO COMBAT COVID-19

However, in the face of the global COVID-19 pandemic, it has been reported that efforts are underway to use drones in a variety of situations, especially over time and not on a large scale. It summarizes what is normally available to the public regarding the use of drones in a situation where there is a serious illness.

10.11.1 Uses of drones

In response to COVID-19, media reports and other available sources have identified three major drone use cases. These include:

1. Unloading and distribution of laboratory samples distribution of medical supplies to reduce travel and reduce exposure in case of infection.
2. Spraying air in public places to decontaminate potentially contaminated areas.
3. Watch out for public places and close and isolated places promptly.

10.11.2 Transportation

There is little public information on the increase in demand and the number of different medical supplies delivered during the epidemic, which makes air delivery operations more efficient. Drones should be viewed differently from other established tourist routes. Drones can be delivered in the context of COVID-19, delivery speed, expanding the transport network to the end, limiting physical contact, and reducing the risk of transmission during delivery.

10.11.3 Aerial spraying

There are many reports within the media of the utilization of drones and disinfectants outside of public places with the potential to unfold the virus. Efforts were created at locations in China, UAE, Spain, South Korea, and alternative countries. Some corporations claim that they were ready to cowl three square kilometers by spraying. However, scientific proof shows that this supplement lacks proof of effectiveness.

10.11.4 Surveillance of the public spaces

Various police and security agencies or organizations (Sierra Leone, Rwanda, China, the United States, Spain, Italy, France, the United Kingdom, India, and others) around the world have used drones to explore public spaces for increased state awareness and segregation enforcement by sending messages, speakers, and tracking down law-abiding citizens. The rental of movies and delivery of voice messages through drones are meant to reduce the likelihood of replies coming into physical touch with potential victims.

Furthermore, additional research organizations began trying drones to follow that enabled hot photographs and installation knowledge. While the use of mass surveillance was common throughout the COVID-19 reaction, numerous human rights advocates have questioned its usage because of potential human rights breaches and even some of these drones have been discontinued.

However, if we fully understand the problem and context, combined with the right drone solution, we can truly provide an archive of advances in the use of this technology, and it should be through the framework of relevant legislation, local capabilities, and the sustainability plan. Image is important to enabling asset purchasing managers to make costly and effective decisions as part of their response to COVID-19. To create a support and approval system, nature needs to focus on a few different but key functions:

- The availability of adequate human and financial resources to have the drone technology available when required; can be done through service contracts or through a local organizational capacity to carry out drone missions. And it takes global movement, health, and service delivery.

- The development of a procurement strategy is based on the selection of the cheapest service that offers quality (service and technology), elegance, stability, compliance, and other key elements.
- The implementation of the drone program cannot be done without local skills and power, so local education and knowledge transfer are key powers. This applies not only to people who claim to be able to perform operations with drones but also to government agencies and end users of this technology in the healthcare sector.
- The use of drones is not possible unless there is a local regulation that allows the safe operation of drones;
- The local sensibility of the communities and participants should take place before and after the implementation of the drone program; public and technical awareness ultimately ensures the social and political acceptance of the place.
- The inclusion of drones in health care should be designed and determined through the design of the existing health system, taking into account the drone problem to be solved, the purpose of this technology, and the clarification of whether drones are suitable for cheaper alternative travel routes.

10.12 ROBOTS DURING COVID-19 SITUATION

COVID-19 can be difficult to deal with in hospitals. People living with the virus should receive treatment, but it is not always easy for healthcare workers to do without putting themselves at risk. Another safe option is to use robots. Robots are used to kill germs in the air and space, answer patient questions, administer medication or food, and control themselves.

Several hospitals in Policlinico Abano, Italy, use *UVD robots* to kill germs in inpatient rooms. The robot can travel to the hospital on its own and kill the virus with UV light.

Pal Robotics has also developed robots that can be used during epidemics. The robot "Ari" can ensure that patients are taking their medication and it can also communicate with patients so that they are not overwhelmed by a lack of community mobilization.

The TIAGo basic robot can be used in two different ways: First, it can be converted into a TIAGo delivery robot that transports food and medication to COVID-19 patients; Second, it can be used as an independent disinfection robot or ADR to kill germs in rooms with UV light.

Another use of the robot can be the remote detection of details of important patient symptoms. Boston Dynamics is currently working on this with its robots and they are called Pal Robots. They are trying to find a way to measure the distance to body temperature, breathing rate, pulse rate, and oxygen supply. Again, this will allow healthcare providers to stay in touch with infectious patients while maintaining high-quality care. Regular monitoring of patients with no risk of infection can be a turning point [14].

This great revolution should be accompanied by a deep understanding of the situation, which includes the perspectives of all interested parties.

10.13 INTERNET OF THINGS BUTTONS

IoT buttons are small and inconspicuous, and they are available from a range of shops using various technologies. IoT buttons may be used for a variety of reasons, such as a button in a patient room that, when pressed, notifies that the room needs to be cleaned. Alternatively, the buttons may serve similar functions, such as a patient call button that requests assistance but does not specify the kind of assistance.

When you press the IoT button, a predefined message is sent over the network (Wi-Fi) to a server that can deliver customized notifications. Notifications can take the form of a regular text message [short message service (SMS), an email (email), or a web page]. Furthermore, alerts may be extended practically endlessly by using application programming interfaces, which are software techniques that allow computer programs to exchange information [15].

10.13.1 Advantages

Traditional hospital notification systems can be devices such as patient call buttons, quick response codes (QR codes), or technology such as information retrieval portals. Electronic medical record (EMR) has also stimulated the development of EMR rules and dashboards for EMR-based notification management. These information systems often require several steps to work or are designed for different tasks. The messages can be customized for specific tasks and can be delivered in a variety of configuration types.

IoT nodes can also be based on different technology platforms. Although the IoT button is a visual tool, activating it can cause an explosive increase in activity in other programs, such as medical records or online search portals. A web interface is also used to measure usage of IoT nodes, unlike other existing hospital advertising systems.

Hospital systems need to understand the existing privacy and security risks when considering using IoT keys. Confidentiality breaches can occur when unwritten messages containing protected health information are collected. Cybersecurity can be compromised when IoT keys are used to access hospital networks or used as a Denial of Service (DDoS) attack. Understanding these risks and developing strategies to successfully minimize them while deploying IoT nodes in a safe manner.

10.14 CONCLUSION

When it comes to identifying an infected patient, IoT looks to be the most straightforward method. This technology aids in the maintenance of internal control concerning period information in health care. COVID-19 serves as a motivation for technical advancement and innovation. To control the spread of the illness, all nations have taken serious action. Modern technology plays an important part in the fight against coronavirus. The method significantly minimizes the effects of coronavirus. Technologies such as AI, cloud computing, the IoT, detection technology, robots, and others help significantly to mitigate the consequences.

Because the COVID-19 batch forecasting system is currently ineffective, there are no practical means for delivering the virus in many nations. As a result, in this paper, we suggest a future low-cost technology for monitoring media interactions and predicting COVID-19. The solution is based on a mobile device with IoT capabilities. Wearable gadgets are gaining popularity and are predicted to aid in the development of applications for both health care and the afterlife, particularly when combined with IoT systems. The technology may have certain disadvantages that offset the program's benefits. Today, as we face the extraordinary effect of COVID-19, entrepreneurs must migrate to the cloud. I believe that cloud absorption will continue to accelerate after COVID-19. However, once the pandemic is behind us and we enter a solid "new normal" (hopefully soon), industry analysts will have to see if institutions are looking at cloud services as a temporary response to this storm.

IoT technology provides emerging economies with the possibility to effectively combat COVID-19 while also accelerating the digital integration of healthcare systems to bridge the essential accessibility, quality, and price gap except for COVID-19, the continued development of the IoT might aid in the prediction of future epidemics. Using statistical methodologies and the advancement of AI and big data, the IoT may be positioned as a critical instrument for quickly transitioning health care from efficient functionality to functional systems.

REFERENCES

1. S. Tyagi, A. Agarwal, and P. Maheshwari, "A conceptual framework for IoT-based healthcare system using cloud computing," in Proceedings of the 2016 6th International Conference-Cloud System and Massive Data Engineering (Confluence), pp. 503–507, Noida, India, January 2016. https://ieeexplore.ieee.org/document/7508172
2. C. Wohlin, "Guidelines for snowballing in systematic literature studies and a replication in software engineering," in Proceedings of the 18th International Conference on Evaluation and Assessment in Software Engineering, Berlin, Germany, March 2014. https://dl.acm.org/doi/abs/10.1145/2601248.2601268

3. P. Kaur, R. Kumar, and M. Kumar, "A healthcare monitoring system using random forest and Internet of Things (IoT)," Multimedia Tools and Applications, vol. 78, no. 14, pp. 19905–19916, 2019. https://link.springer.com/article/10.1007%2Fs11042-019-7327-8
4. S. Mohapatra, S. Mohanty, and S. Mohanty, "Smart healthcare: an approach for ubiquitous healthcare management using IoT," in Big Data Analytics for Intelligent Healthcare Management, pp. 175–196, Elsevier, Amsterdam, Netherlands, 2019. www.sciencedirect.com/science/article/pii/B9780128181461000076
5. P. J. Nachankar, "IoT in agriculture," deciding, vol. 1, no. 3, 2018. https://scholar.google.com/scholar_lookup?title=IOT%20in%20agriculture&author=P.%20J.%20Nachankar&publication_year=2018
6. Cloud computing systems and applications in healthcare. IGI Global, Hershey, PA2016. www.google.co.in/books/edition/_/f4XvDAAAQBAJ?hl=en&gbpv=1&pg=PR1
7. ZYMR Blog, "Cloud Computing in Healthcare – Advantages and Benefits of Cloud Computing" www.zymr.com/5-key-benefits-of-cloud-computing-in-healthcare-industry/
8. IEEE COMSOC, "Smart IoT Solutions for Combating COVID-19 Pandemic." September 2020. www.comsoc.org/publications/magazines/ieee-internet-things-magazine/cfp/smart-iot-solutions-combating-covid-19
9. S. Kumar, P. Tiwari, and M. Zimbler, "Internet of things may be a revolutionary approach for future technology enhancement: a review," Journal of Big Data, vol. 6, 2019:111.10.1186/s40537-019-0268-2. https://journalofbigdata.springeropen.com/articles/10.1186/s40537-019-0268-2
10. S. Zheng, Y. L. Qian, Z. P. Xiang, L. H. Bo, F. Liu, and Z. R Sheng, "Recommendations and guidance for providing pharmaceutical care services during COVID-19 pandemic: a China perspective," Research in Social and Administrative Pharmacy, vol. 17, no. 1, pp. 1819–1824, 2020. www.ncbi.nlm.nih.gov/pmc/articles/PMC7102520/
11. A. Abbasian Ardakani, U. R. Acharya, S. Habibollahi, and A. Mohammadi, "COVIDiag: a clinical CAD system to diagnose COVID-19 pneumonia supported CT findings," European Radiology, vol. 31, 2020 www.ncbi.nlm.nih.gov/pmc/articles/PMC7395802/https://pubmed.ncbi.nlm.nih.gov/32740817/
12. Fierce Electronics: "The role of IoT sensors within the COVID-19 fight" www.fierceelectronics.com/electronics/role-iot-sensors-covid-19
13. "Industry 4.0 technologies and their applications in fighting COVID-19 pandemic." https://pubmed.ncbi.nlm.nih.gov/32344370/
14. "The Clever use of ROBOTS during COVID-19": https://hospitalityinsights.ehl.edu/robots-during-covid-19
15. https://onlinelibrary.wiley.com/doi/full/10.1002/ima.22552 "Future IoT tools for COVID-19 contact tracing and prediction: A review of the state-of-the-science" February 2021.

Index